T0227867

ENVIRONMENTAL IMPACTS OF WASTE PAPER RECYCLING

Yrjö Virtanen and Sten Nilsson

IIASA
International Institute for Applied Systems Analysis

Earthscan Publications Ltd, London

from Routledge

The views and opinions in this document do not necessarily represent those of IIASA or the organisations that support its work.

First published 1993 by
Earthscan Publications Limited

Earthscan

2 Park Square, Milton Park, Abingdon, Oxon OX14 4RN
Simultaneously published in the USA and Canada by Earthscan
711 Third Avenue, New York, NY 10017

First issued in hardback 2017

Earthscan is an imprint of the Taylor & Francis Group, an informa business

British Library Cataloguing-in-Publication Data

A catalogue record for this book is available from the British Library

ISBN 13: 978-1-138-42296-4 (hbk)
ISBN 13: 978-1-85383-160-7 (pbk)

Earthscan publishes in association with WWF-UK and the International Institute for Enviroment and Development.

Contents

Preface

Our concern for conservation of our natural resources and about the deleterious effects on the environment of disposal of waste products is increasingly reflected in proposed legislation aimed at reducing waste. The preferred technique is recycling of waste products.

While laudable in its objective, a narrow focus on recycling is limited, and can result in unexpected effects that can at least partially offset the expected benefits. This is particularly true of paper for at least three basic reasons. First, paper is a major component – about 35% – of household waste volume but sites for paper disposal do not coincide with the sites for pulp and paper production. Second, unlike most waste, paper has a very high energy content. And third, unlike coal or oil, paper is a renewable resource, and in Europe is produced mostly from forests managed on sustainable principles.

This book summarizes a feasibility study of large-scale paper recycling in Europe. The study investigated the entire production and disposal process using a "life-cycle" methodology and database (Lübkert *et al.*, 1991) developed at IIASA. In addition, the feasibility study also used data and results produced by IIASA's Forest Resources Project (Nilsson *et al.*, 1992).

The conclusions of the study, while too preliminary to permit solid quantitative comparisons, indicate that paper recycling has economic and environmental advantages. However, because paper is a renewable resource and has a high energy recovery content, there is the possibility of energy recovery as a suitable option. A balanced mixture of paper recycling and energy recovery seems to be the most suitable solution since recycling minimizes the use of some resources and emissions, and energy recovery minimizes the overall use of fossil fuels. A number of important questions remain that must be further investigated before large-scale programs for increased recycling of paper products are introduced, since the environmental impacts are shown to depend strongly on how extensive and how selective recycling is. The extent of recycling will influence the role of energy recovery and the possibilities of carrying out the necessary silvicultural operations. Thinnings and other silvicultural operations are crucial for the future vitality of the European forests.

Sten Nilsson
Leader
Forest and Climate Change Project, IIASA

Summary

We live in a throwaway society, and much of what we throw into Europe's growing rubbish mountain is paper: paper makes up about 35% of total household waste volume. Driven by the anxieties of environmentally concerned citizens, many countries have introduced legislation designed to reduce waste very quickly. Among the main arguments behind the popularity of planning materials recovery from the starting point of "closed loop recycling" is the general belief in less consumption of resources, less energy consumption, cheaper production costs, and an overall reduction of environmental load through recycling.

Obviously, recycling is a means of reducing waste streams and, accordingly, reducing the demands for waste-treatment capacity. It is perhaps less obvious that increased recycling may also actually *increase* the consumption of nonrenewable resources. Hence, the objective of an efficient material production and recycling scheme should not be to recycle *per se*, but rather should be to minimize the resource utilization and emissions of all streams of materials in the production cycle, from "cradle" to "grave."

To identify such optimal schemes, it is necessary to consider many different alternative arrangements for material management, because the advantage gained in one respect might easily be lost in another.

This was the general objective of an IIASA feasibility study of recycling paper products in Western Europe. The specific objectives of the study were to evaluate the applicability of a life-cycle approach to paper recycling, to provide new insights into the complexity of introducing large-scale recycling into existing production and distribution systems, and to broaden the debate with new arguments. To identify the outer boundaries for paper recycling in Western Europe and its implications, three different scenarios were

Table S.1. Results of calculations for conditions prevailing in the late 1980s and for increased recycling rates.

Energy consumption		*Emissions – Water*	
Electric power	Decreased	TSS	Increased
Heat and steam	Decreased	BOD	Increased
Fossil fuels	Increased	COD	Decreased
Nonrenewable primary		AOX	Decreased
energy sources	Increased		
Renewable primary		*Materials*	
energy sources	Decreased	Raw materials for	
		pulp and paper	
		production	
Emissions – Air		(other than wood)	Decreased
SO_2	Increased	Wood consumption	Decreased
NO_x	Increased		
CH_4	Decreased	*Waste production*	
Gross CO_2	Decreased	Gross solid waste	Decreased
CO	Decreased	Net solid waste	Increased
Net CO_2	Increased		
	(or decreased	*Forest management*	
	fixation)	Intensity	Decreased

selected: one in which recycling was used to the maximum extent feasible; one in which recycling was used selectively; and a third where recycling was not used at all, but rather waste paper was used as an energy source. The scenarios may not represent realistic future recycling strategies, but they serve the primary purpose of demonstrating the scope of sensitivity of the environmental impacts to different recycling strategies. It should also be pointed out that while we feel that the conclusions of the study correctly indicate trends, they are too preliminary to permit solid quantitative comparisons.

The results in *Table S.1* clearly illustrate that identification of optimal material production and recycling schemes, from an environmental point of view, is much more complex than what is included in the current debate. Recycling of paper in Western Europe clearly has economic and environmental advantages, but while maximum recycling reduces demands, maximum recycling also *increases* consumption of fossil fuels and *increases* emissions such as SO_2, NO_x, and net CO_2.

The maximum recycling scenario shows a forest utilization considerably below that estimated as a sustainable level for Western Europe. Underutilization of forest resources may lead to unsatisfactory economic conditions for the necessary forest management, resulting in fewer vital forests and higher vulnerability to natural stress and air pollutants.

The results also indicate that a balanced mixture of recycling and energy recovery seems to be a suitable solution, since recycling minimizes the use of certain resources and reduces emissions, while energy recovery minimizes the overall use of fossil fuels. The appropriate balance may vary from country to country in Western Europe.

The results clearly show that several important questions must be investigated further before large-scale programs for the increased recycling of paper products are introduced with the objective of improving the environment.

Acknowledgments

In 1986 the IIASA Forest Study began to address the question of long-term development of European forest resources. The current feasibility study on the environmental impacts of paper recycling in Western Europe has been carried out by the IIASA Forest Study in close collaboration with experts from industry and research organizations. We would like to thank all of our collaborators. We would especially like to thank Barbara Lübkert-Alcamo, Vesa Junttila, and Börje Kyrklund for their substantial effort in collecting the material and shaping the contents of this book.

We would also like to express our sincere gratitude to Muriel Weinreich and Cynthia Ramirez, IIASA, who assisted us with the different versions of the manuscript.

Chapter 1

Background, Objectives, and Methodology

There is a growing awareness of the need to radically decrease waste streams from production and consumption processes. This awareness has not only brought about the implementation of improvements in processes but has also led to increased circulation of materials. Unfortunately, industry has not always been able to make use of all reusable materials available; on the other hand, collection of the materials for reuse has not been as efficient as was estimated or expected. This has led to increasing frustration among both consumers and industry toward policy makers. To a large extent, this dilemma has arisen from the incompatibility between the goals of policy makers and the actual possibilities of rapid changes in production processes and consumer behavior. This incompatibility could only be avoided by setting more realistic goals for the reduction of waste streams, thereby reducing the excess costs resulting from inefficient policies.

The sheer volume of waste, particularly solid waste, complicated by limited waste-management resources, has led to changes in consumer behavior, to the introduction of legislation intended to reduce waste volume, and to great improvements in industrial technology. For example, during the past 20 years, despite increased production, the total wastewater discharge from paper and pulp production in some Western European countries has been halved, and the total biological oxygen demand (BOD) load reduced to one-third its former value (National Board of Waters and the Environment, Finland).

Driven by the concerns of environmentally concerned citizens, many communities and countries have introduced legislation designed to reduce waste very quickly. For example, the British target is a 50% overall recovery of recyclable household waste by the year 2000 (UK Environmental Protection Act), the German target is 80% by July 1995 (German Dual System Regulations), and the EC target is 60% by the year 1996 (Draft Directive on Packaging). In the Netherlands, industry has undertaken to reuse at least 60% of material, so that by 1995 the amount of packaging currently going to landfills will be reduced by 60% (Environmental News, 1991).

Targets for materials recovery are set, in particular, to provide substitutes for primary materials in the manufacture of goods. But recovery of materials as substitutes for fuels in energy production is currently often excluded from political recovery plans, even though the development of incineration technology and the reduction of heavy-metal, chlorine, and other contaminants in wastes might be an essential future strategic alternative.

One of the main arguments behind the popularity of planning materials recovery from the starting point of *closed loop recycling* is the general belief in the overall reduction of environmental load through recycling. Obviously, recycling is a means of reducing waste streams and, accordingly, reducing the demands for waste-treatment capacity. But, on the other hand, recycling may have the opposite effect of increasing demand for resources. The facilities and activities required for managing recycling, and the need to add material to compensate for quality degradation, consume energy and materials.

Paper differs from other basic material in Western Europe (and also in the rest of the world) in several fundamental ways. First, paper comes from a renewable source. Second, it possesses a high energy potential. Third, because of geo-climatic circumstances, the centers of consumption and the sources of raw material are far apart. Because of renewability, the application of the principle of sustainability to paper should be focused on managing the wood balances rather than overall minimization of the use of raw wood material. In addition, the energy potential of paper should be taken into account as an alternative to nonrenewable energy sources. Utilizing the heat potential of waste paper represents an essential way of both saving nonrenewable resources and minimizing solid wastes.

The objective of an efficient material production and recycling scheme should be to minimize the resource utilization and emissions of all streams of materials from *cradle* to *grave*. When searching for such an optimal scheme, it is necessary to consider many alternative arrangements for material management, because the advantage gained in one respect might easily be lost in another. One of the cornerstones of such considerations should be based on an objective and comprehensive impact inventory of the alternatives, rather than intuition.

The objectives of this feasibility study on recycling paper products in Western Europe were to demonstrate and evaluate the applicability of the life-cycle approach and methodology to the paper-recycling problem and, from this starting point, become involved in the debate on material management strategies and the concept of sustainability: to present reasons for questioning the arguments of the public debate about recycling of paper products, rather than provide evidence verifying or disproving them. The results of this study should shed light upon the following questions:

- *Would the maximum use of recycled fiber in all paper qualities reduce environmental impacts?*
- *Could lesser impacts be achieved with a selective recycling strategy?*
- *What would be the effects in the case with no recycling and maximum energy recovery?*
- *What would be the potential impacts of different paper-recycling strategies on wood balances and on silviculture in Western Europe?*

The present study approaches these questions by investigating the differences in the environmental loads and raw material demands from production and consumption, at the general level, as aggregated in two regions: Scandinavia and Central Europe. For the sake of simplicity, Scandinavia comprises Sweden and Finland and Central Europe comprises Germany, France, Italy, the Netherlands, the UK, and Austria. The assumption is that the imbalances between supply and demand of waste paper in these regions are balanced either by exports or imports or by directing surplus supply to the waste-handling processes. The latter approach of directing part of the surplus supply (e.g., in the low-quality grades) to waste handling does not necessarily mean increased environmental burden alone, since the energy content of the fibers can be recovered by incinerating the recycled paper.

At this stage, the preliminary nature of the data used for inventories and the lack of refinement in the model of the recycling system do not allow solid quantitative analyses, evaluations, or comparisons. An essential objective of a planned full-scale study at IIASA on recycling of paper products in Western Europe is to collect new data and refine the model to such an extent that even quantitative evaluations will be relevant and justified.

The methodology used in the IIASA feasibility study on recycling paper products in Western Europe is life-cycle analysis (LCA). The principle of life-cycle analysis implies that products, activities, or even entire economic sectors are analyzed from an end-use perspective. The life-cycle approach makes it possible to quantify the cumulative impacts that a product generates from the point where materials and energies for this product are extracted from nature up to either a certain point in the product's life-cycle or, in the most complete case, the final disposal of the wastes (that is, when it is returned to nature). The processes that the emissions and wastes undergo in nature should be included, but, at present, they are disregarded in the analyses because of their complexity.

Life-cycle analysis has its roots as far back as the early 1960s. At the World Energy Conference in 1963, Harold Smith published a report on the cumulative energy requirements for the production of chemical intermediates. In the late 1960s and early 1970s, several researchers undertook global modeling studies in which they attempted to predict how changes in population would affect the world's total mineral and energy resources (Meadows *et al.*, 1972; Mesarovic and Pestel, 1974). During the period of major world oil crises in the mid- and late 1970s, the United States commissioned about a dozen major "fuel cycle" studies to estimate the costs and benefits of alternative energy systems. Later, similar studies were commissioned by both the US and British governments on a wide range of industrial systems. In 1985, the Commission of the European Communities introduced a Liquid Food Container Directive (CEC, 1985) which charged countries with monitoring the raw material and energy consumption, as well as the amounts, of the solid waste they generated. As concern about global air and water pollution problems increased, these emissions were then also routinely added to energy, raw material, and solid waste considerations.

In the traditional approach of environmental impact analyses, the industries or industrial sectors themselves are studied and their impacts are implicitly taken as representative of the products. Although in most cases the traditional approach gives a rough estimate of the product's impacts, it does not identify all sources of pollution associated with the product and, in many cases, may not even identify the largest or main sources of environmental degradation. For this reason, the traditional approach is not comprehensive enough to identify all possible strategies for reducing a product's environmental impacts.

Although life-cycle analyses identify the amounts of known pollutants from the production systems studied, they cannot, and should not, without further interpretation, be used to compare the environmental toxicity of these production systems. One can readily compare the amount of identical pollutants produced by different production systems, but one cannot deduce from life-cycle analyses whether or not one specific pollutant, or group of pollutants, is more harmful to the environment than another.

Another potential weakness of life-cycle studies is the tremendous amount of data required. It is extremely difficult to document clearly and understandably all data and assumptions that go into the final results of any life-cycle analysis. If different studies of the same products, however, come up with different final results, one should be able to trace these discrepancies back to the different data and assumptions made. This is only possible if all sources of data are accessible and well documented. Life-cycle analysis is an inventory method and, as a result, generates a long list of substances that are either:

- *produced* by the system studied as either *useful products* or wastes discharged into the environment; or
- *consumed* by the system studied in either *material* or energy form.

To use these results for policy decisions, this list of substances needs to be interpreted and, in general, reduced to a limited number of factors. Some researchers consider this step to be part of the life-cycle analysis itself, whereas others refer to it as the *ecoprofile analysis*. This reduction can be done by:

- *using* one or several of the individual substances as *indicator(s)* of the overall impacts; or
- *aggregating* this list into a *limited number of quantities*, such as total material requirements, total air emissions, total water emissions, total solid waste discharges, and total environmental costs.

Both methods involve value judgments. In the first case, representative indicators have to be selected; these almost always depend on the goals set by the person studying the respective production system. For example, if the main objective is to reduce the amounts of solid waste generated, the representative indicator(s) will be different than if the goal is to limit the release of toxic quantities into the environment. In the second case, data for different substances have to be added together into one quantity, or several, but still meaningful, quantities. This requires a value judgment on the comparability of these various substances and on their relative harm to the environment. Even environmental costs can be seen as a subjective issue, although they could also be approached, in part, from the techno-economical aspect of *reduction* costs. Therefore, when discussing life-cycle analysis it is very important to differentiate between an inventory and quantification of the impacts and an interpretation of results in order to answer policy questions.

Life-cycle analyses are used to quantify and compare stresses on the environment caused by alternative products or by different production systems and technologies making the same products. The method can also be used to compare the impacts of entire industries, economic sectors, and even total national economies. The information can then be used in an ecoprofile analysis to help address a number of technical and political issues in several areas, such as the following:

- *Comprehensive environmental impact assessments* – The life-cycle impacts generated by a certain product can be studied and compared with the impacts of other products.
- *Environmental labels* – LCA can furnish a quantitative basis for awarding labels for environmentally benign products (ecolabels).
- *Assessment of industrial processes' efficiencies* – The information can be used to calculate energy and material usage efficiencies

within a given economic sector or activity and identify possible areas where improvements can be made.

- *Evaluation of policy alternatives to minimize environmental impacts* – One can assess the impacts of possible alternative environmental regulations through the analysis of different scenarios in order to find the best regulation.
- *Comparison of environmental performance* – The environmental performance in certain economic sectors of different countries can be compared.
- *International negotiations on environmental policies* – LCA can be used to assess and compare the systems' efficiencies for different geographic regions or countries in order to detect potentials for improvements.
- *Optimization of policies for the ecorestructuring of economies* – LCA provides a tool for evaluating ways of restructuring a national energy system in an environmentally sound way.

The IIASA feasibility study on recycling paper products in Western Europe, however, has been limited to demonstrating only the general justification for and possibilities of LCA methodology. The impact inventory is based on preliminary data and rather rough assumptions about the recycling system. Accomplishing a comprehensive assessment is a major task that requires resources far beyond those available for this study. Nevertheless, the inventory results from this study offer a factual basis for introducing new arguments into the debate on recycling paper products.

Chapter 2

Problems of Paper Recycling in Western Europe

To date, approximately 30% of fibers are recycled worldwide, with some countries reaching 50%. In the 1980s, the use of recycled fibers increased rapidly. In 1986 the overall waste paper use (raw material input per production output) in Western Europe was only about 35%, which corresponded to about 30% furnish share in the pulp mixture. Global use at the end of the 1980s was approximately 75 million tonnes of recycled fiber per year, and it has been estimated that by the year 2000 this could increase to 130 million tonnes per year (Jaakko Pöyry, 1990).

The reasons for the low share of waste paper use in 1986 are obvious. The quality of the paper made from recycled fibers was often considered insufficient for newsprint and other printing paper uses. However, about half of the total paper and board produced in Western Europe in 1986 was for these uses. As mentioned earlier, in 1986 less than 30% of the total fibers used in newsprint were recycled fibers. In addition, recycled fibers were hardly used at all in high-quality wood-free grades. On average, waste paper raw material input was used in less than 10% of all printing paper. The average raw material furnish in Central Europe was about 12%; Scandinavian production, which represented about 40% of the total Western European supply, was firmly based on primary fiber. Thus, paper recycling in Western Europe could best be characterized as

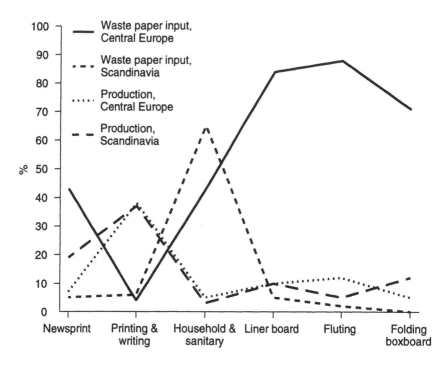

Figure 2.1. Paper and board production and waste paper (raw material) furnish profiles in Central Europe and Scandinavia in 1986 (Benn Publications, 1987; OECD, 1989; FAO, 1991b).

typically limited to one reuse cycle of high-grade waste paper to lower-quality papers.

At present, recycled fiber is widely utilized in Central European countries for corrugated board materials, liner boards, and fluting (*Figure 2.1*). Acceptable qualities are produced with an 85% to 90% recycled fiber furnish; the average utilization is about 70%. The use of recycled fiber has increased in folding boxboards and in sanitary and household papers, where even a 95% to 100% share of recycled fiber furnish has been attained in actual production.

Currently, the highest percentages of recycled fiber are often gained by means of selective waste paper collection of easy-to-use raw material for fiber recovery. Today, the grades utilized are mostly lightly printed or unprinted; no de-inking or separation of non-fibrous materials is necessary. It should be noted, however, that in Europe the resources of such materials are fairly limited. In the case of large-scale recycling of fibers, lower grades of waste paper having higher demands on utilities and materials, and higher emissions from processing, should also be included. When planning the collection and recycling of waste paper, it is also necessary to consider the different qualities of the paper; high-grade waste paper can be recirculated and used for producing lower-quality paper, but the reuse of low-grade waste paper is extremely limited. Also, paper cannot be recirculated without additional primary fibers because a certain quality degradation will occur at each reuse cycle. Therefore, separating waste paper into different qualities would facilitate the use of recycled fibers and here, for certain qualities, the best use would be to recover the energy content of the fibers.

A general problem with paper recycling in Western Europe is that pulp and paper production and consumption sites are geographically separated. Scandinavian countries are major producers but only minor consumers. Consequently, the question of availability and logistics of waste paper recycling enters the equation.

The imbalance of production and consumption in Scandinavia is evident for all paper and board qualities, except for household and sanitary papers typically produced for local markets because of relatively high transportation costs per product volume (*Figure 2.2*). The Central European production–consumption balance is negative. Because of the large total volumes of production and consumption the differences are relatively small for all qualities except newsprint.

In 1986 Sweden and Finland together had a 9.4-million-tonne positive trade balance (production minus own consumption) in paper and board products, while Central Europe (Germany, France, Italy, the Netherlands, the UK, and Austria) had a 6-million-tonne negative trade balance. The total trade balance for Scandinavia and Central Europe was +3.4 million tonnes, representing about 10% of total production. The most significant export areas were the USA and Asia (collectively 2.6 million tonnes, OECD, 1989).

From this rough balance two conclusions could be drawn: (1) an overall, general recycling of paper fibers in Western Europe, under the current production capacity distribution, evidently means increased long-distance transport and (2) part of the reuse fiber potential is lost because of intercontinental export. To even out the fiber potential in Western Europe would mean about 10 billion

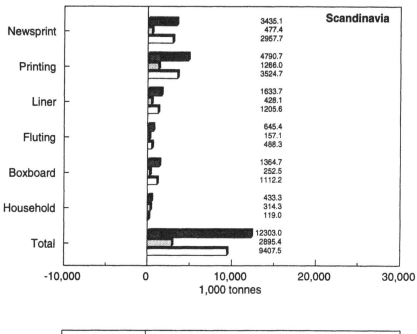

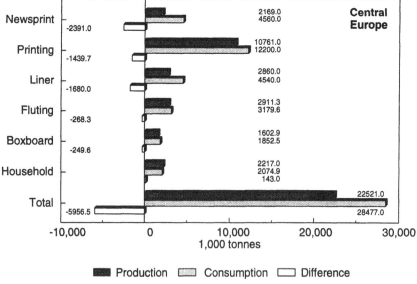

Figure 2.2. Production–consumption balance of paper and board for Scandinavia (Sweden and Finland) and Central Europe (Germany, France, Italy, the Netherlands, the UK, and Austria). Source: OECD, 1989.

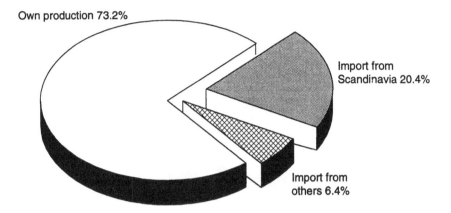

Figure 2.3. Central European paper and board supply (OECD, 1989; FAO, 1990).

tonne-kilometer transport output by rail and ship between Scandinavia and Central Europe. About 10% of the reuse fiber potential produced would thus be irreversibly lost, if no intercontinental transport of waste paper is achieved.

In 1986, about 27% of the Central European paper supply was based on imports, mainly from Scandinavia (*Figure 2.3*). If this trend continues, transportation networks between producers and consumers should be built up. A production facility relying mostly on recycled materials could, however, be placed near the main consumption centers instead of the main sources of primary raw wood material. These considerations indicate one degree of freedom in the search for optimal production and recycling structures.

A scenario with large-scale paper recycling immediately includes the option of regional (and international) restructuring of the existing pulp and paper industry. In addition to the paper import, about 30% of the wood pulp used in Central Europe was imported, again to large extent from Scandinavia (*Figure 2.4*). Consequently, large-scale paper recycling in Western Europe would have a radical impact on the current primary fiber market. For this reason and because of the large, new investments needed for reused pulp production, and also the large, existing investments in primary pulp production facilities, the changeover to large-scale paper recycling may take many years and may encounter economic and political difficulties.

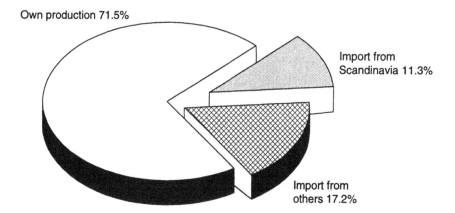

Figure 2.4. Central European wood pulp supply (OECD, 1989; FAO, 1990).

The restructuring aspect of the current industry has been omitted in this study due to the complexity of the problem, even though it has potential significance for the overall environmental impacts. Primary pulp production, for instance, could be expected to become, for the most part, separated from paper production. This would mean increased use of market pulp and, accordingly, increased energy consumption. In this IIASA feasibility study, the production structure is assumed to remain as it is for the different scenarios produced. The problem of restructuring is, however, one of the topics which will be addressed in the planned full-scale study on recycling of paper products in Western Europe.

Large-scale paper recycling could also have a fundamental impact on the European wood balances and, thereby, on the possibilities of practicing sound forest management in the region. About 80% of the raw wood material used in the late 1980s in Western Europe's pulp industry was in the form of pulp-logs (UN, 1991), the main part of which came from thinnings. If increased recycling were to cause the raw wood demand to sink significantly under the biological harvest potential, it may be difficult to maintain silvicultural standards and the vitality of the forest resources. It can be argued, on the other hand, that the thinnings not used by the pulp industry could be used by the board industry. This is, however, unlikely to occur because the board industry's paying capacity for the pulp-logs is much lower in comparison with the pulp industry.

An interesting issue related to paper-recycling strategies is the economic potential of waste paper for energy production. For some national economies it could be more profitable to use waste paper for energy production than to use it as raw material for domestic paper production or to export it. Depending on the proportion of wood fibers, the heat content of waste paper is from 14 to 17 megajoules per kilogram (MJ/kg), which makes 1 tonne of waste paper equal to about 0.4 tonnes of oil. We cannot, at present, generalize the relative advantage of recovering waste paper for its energy versus its fiber value, since the most economical utilization of waste paper depends on which fuels it would replace and the type of paper produced from it. Such calculations can be made in the planned full-scale study on recycling of paper. However, it should be pointed out that waste paper for energy would be a clean source of energy if some of the chemicals and heavy metals used in the pulp, paper and board, and printing process were replaced with other, more benign materials.

Using waste paper for energy production provides benefits which should be compared in larger-scale fiber recycling with the energy input necessary for its collection and re-pulping. The interactions between different economic sectors may also introduce problems if the heat value potential of household waste decreases because of waste paper recovery. A conversion of the benefit of using recycled fibers to an equivalent amount of unused oil needs to be considered. Ignoring these considerations will give a distorted view to policy makers and consumers of the costs and benefits of recycling. When considering paper products and the basic wood material, it should be noted that they represent one of the most used renewable material and energy source of the world. An efficient forest management program is necessary, however, to avoid deforestation and corresponding impacts on expected global climate changes. Including new forest areas in a better management program may improve the effects on the fixation of carbon dioxide and the release of methane to the atmosphere.

There are also technical limitations to large-scale recycling. Fibers degenerate each time they are reused, which limits the number of reuse cycles to between three and five. On the other hand, a sizable percentage (20% to 25%) of paper is used for purposes that make recycling impossible or infeasible. Papers belonging to this group are, typically, sanitary papers and food parchment papers, for hygienic reasons, and construction and archive papers from a life-span point of view. Another limitation to using recycled fibers are also the mixture of brown and white, respectively, chemical and mechanical, fibers in the collected waste paper. In addition, different kinds of contaminants in the waste paper mixture are sometimes difficult to remove in the recycling process.

To meet the physical property demands of the actual products and their manufacture, there are several ways of modifying and furnishing of the base paper or board. Normally, recycled fibers have partly lost the properties of primary fibers. For example, re-pulping causes fibers to shorten, which, in turn, implies reduced strength and moisture properties. Depending on the average age of the recycled fibers used in the furnish, the thickness of the base paper or board has to be increased to maintain the required strength and scoring level. Alternatively, undersized fibers could be screened out of the stock; however, because other fiber properties are also involved (such as strength, scoring, and folding), screening alone would not necessarily guarantee the adequate properties. It would also mean less yield of fibers though the most aged material is always screened away or dissolved. The common solution to the problem is to add primary fibers to the furnish by either mixing them with reused fibers or assembling the paper as layers of different qualities of fibers. The specific weight of papers in the cases where recycled fiber is used tends to be 5% to 15% higher than in the cases where only primary fibers are used (Judt, 1991).

The environmental impact of recycling on greenhouse-gas emissions is an important issue to be considered. What are the consequences if wood balances (the yearly increment in relation to the yearly harvest of wood) become largely positive? How do greenhouse-gas emissions from the natural cycle of the surplus wood compare with emissions from incineration? These questions could not be studied completely in the current feasibility study but are to be evaluated in the planned full-scale study.

Chapter 3

Recycling Scenarios and Impacts Studied

To identify the outer boundaries for paper recycling in Western Europe and its implications three different scenarios were selected under the conditions and limitations discussed in Chapter 2. The three scenarios have been selected to represent the two possible extremes of recycling of paper products and a reasonable alternative, illustrating the flexibility of the recycling strategies and their environmental impacts. These scenarios may not represent realistic future visions, but they serve the primary purpose of demonstrating the sensitivity of the environmental impacts of different recycling strategies. The scenarios are as follows:

(M) Maximum and equal share of recycled fiber for all paper and board qualities.

(S) Maximum selective use of recycled fiber, with easy-to-use waste paper grades and geographical availability.

(Z) No recycling and maximum energy recovery of the fibers.

In the *maximum recycling scenario (M)*, a certain percentage of the waste paper is assumed to be collected and used as raw material for recycled fiber forming an equal relative share of the furnish in all paper and board qualities. The collection percentage is assumed to be the highest possible (90% for most of the qualities). The assumed overall average furnish share for the reused pulp is 56% for all paper and board grades. Although, such an average furnish share is not possible today due to technological constraints, the objective of this scenario is to illustrate one of the extremes of paper recycling. The waste paper supply is assumed to be based on Western European domestic markets. Waste paper imports from other parts of the world are assumed to be minimal. Re-pulping technologies are assumed to be available, with de-inking, separation of non-fibrous materials, refining, and screening of the recycled fiber carried out to such an extent that it matches the fiber property demands of each paper quality. The disposed part of the waste paper is assumed to be incinerated (26%) or shipped to landfills (74%).

In the *selective recycling scenario (S)*, the criteria for the waste paper grades and the respective paper qualities they are used for are the simplicity of fiber recovery and the availability of waste paper. Simplicity of recovery means that there are existing and feasible technologies for fiber recovery and that there would be no significant decrease in the quality of the papers caused by furnishing them with recycled fiber. The availability factor means that the waste paper could be collected inside a reasonably compact area (within a range of 150 km to 200 km). Long-distance transportation of waste paper is assumed to be minimized. The disposed part of the waste paper is assumed to be incinerated or shipped to landfills in the same proportion as the previous scenario (26% and 74% respectively). The overall furnish share for the reused pulp is 35% for all paper and board grades, which is about 5% higher per unit than the rate estimated in 1986.

The *zero recycling scenario (Z)* represents an alternative strategy with the maximum energy recovery of paper fibers. In general, energy use is the most important factor when considering environmental impacts. The use of waste paper for energy production reduces solid-waste streams and saves fuel. It might also reduce overall direct emissions from energy production into the air and water, because of the relatively high emission protection level of incinerators, depending on the mixture of the substituted fuel. Energy production that is assumed to be substituted for incinerator output in this scenario is assumed to follow the current average profile for Western Europe. The disposal pattern for waste paper is assumed to be changed to 100% incineration, that is, all the waste paper is assumed to be burned.

For all scenarios, 35% of the heat recovered by waste paper incineration is assumed to be converted into electricity and the rest (65%) to be removed as low-temperature waste heat. Currently, in many European countries fossil-fuel-based district heating may be partially replaced by coproduced heat. However, such possibilities of utilizing the low-temperature heat from waste paper incineration could not be studied in this feasibility study due to limited resources. Because of their anticipated importance to the overall environmental impacts, they are intended to be studied and taken into account in the planned full-scale study.

The data used for the inventories are basically the same for all scenarios. The majority of the data from the basic production, transportation, energy conversion, and waste-management sectors comes from IIASA's IDEA database (Lübkert *et al.*, 1991). These data were collected between 1990 and 1991 from industrial and official statistical sources and, in practice, represent the current situation in Western Europe. The remaining data have been obtained from the latest available OECD statistics (statistical year 1986), and processing data are from individual industrial sources. Parts of the scenarios, however, involve assumptions that do not correspond to the present technology sphere; the data for these parts of the scenarios are estimates, but they are still realistic for the needs of the primary objectives of this feasibility study.

For each of the three scenarios the following environmental impact indicators were calculated.

- *Energy demands:* electric power and heat.
- *Nonrenewable fuel demands:* hard coal, brown coal, derived coal, middle distillate, light fuel oil, heavy fuel oil, and natural gas.
- *Raw material demands:* common raw materials.
- *Consumption of primary energy sources:* coal seam, crude oil, hydropower, nuclear fuel, crude natural gas, and biomass of trees.
- *Air emissions:* CH_4, SO_2, NO_x, CO, and CO_2.
- *Water emissions:* total suspended solids (TSSs), biological oxygen demand (BOD), chemical oxygen demand (COD), and chlorinated organic compounds (AOXs).
- *Solid wastes:* gross municipal wastes (output from the life cycle) and terminal municipal wastes (in landfills).

One criterion for selecting these indicators was that they are the most commonly referenced factors in the political discussion on environmental impacts. In addition, they are most consistently represented in the IDEA database (Lübkert *et al.*, 1991).

Chapter 4

Paper-Recycling System

There are various property requirements of papers and fibers, depending on the characteristics of the actual product the paper or board is used for. Basically, there are two general product groups for which paper and board are used: graphic products, such as newspapers, books, advertisements, and copies; and paper and board products, including packaging papers and boards, household and sanitary papers, and construction papers.

Inside these two groups there are various subgroups with specific property requirements, such as brightness, strength, grammage (area weight), thickness, and moisture re-formation. For large newspapers, for instance, it is essential to have paper with minimum thickness yet with maximum printing properties.

Graphic products represent almost half of the total paper and board consumption in Western Europe. The other half is used for different paper products, with about 35% used for corrugated board and solid boxboard.

The cycle of paper begins at the biomass formation of trees (*Figure 4.1*). Atmospheric carbon dioxide and water from soil are combined in photosynthesis to form glucose, the material basis for the trees' growth. Carbon remains fully in the biomass, but oxygen is released back to the atmosphere. Carbon dioxide fixation is 0.7 to 0.9 tonnes for each tonne of biomass. Many of the trees used for pulp production in Western Europe are grown in cultivated forests.

Paper fiber starts its way toward consumers at harvesting. An increasing part of pulp-logs is gained as a by-product of forest thinning. Today a major part of the felling and limbing is still based on

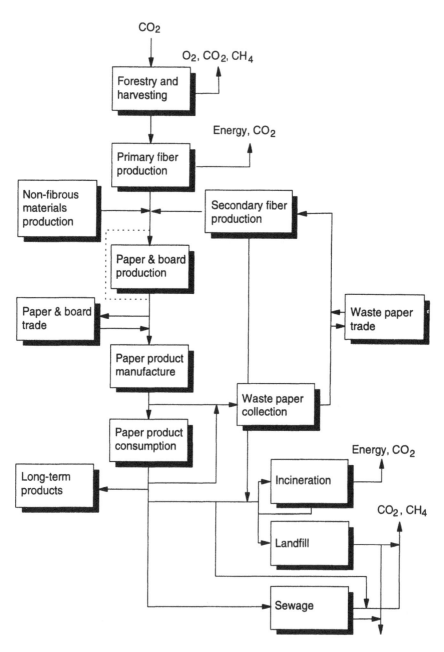

Figure 4.1. The main flows of the paper cycle.

manual technologies even though highly automated and powerful harvesting machines are occupying more space in the harvesting market. The part of the wood biomass in logs leaving forests for pulp mills is about 65% of the total biomass of the trees (National Board of Forests, Finland). The rest, made up of roots, branches, and leaves, normally remains in the forests where it decomposes into methane and carbon dioxide in from five to ten years.

Primary wood fiber is extracted from logs in pulp mills. There are two principal families of technologies for the extraction: chemical and mechanical. Both families comprise several different technologies. There are also technologies utilizing a combination of these principal technologies characterized as semi-chemical or chemi-mechanical. These, however, represent a small minority of the overall pulp production capacity.

The two main technologies differ from each other in several ways. First, in chemical pulping, lignin and other impurities are separated from the cellulose fibers by a chemical-cooking treatment. In mechanical processing, practically all the wood material, except bark, is utilized for pulp. Second, chemical processing is practically energy self-sufficient due to utilization of the heat potential of extracted lignin and bark. Mechanical pulping is fully dependent on external energy. In addition, the specific electric energy demand for chemical pulping is roughly one-third of that for mechanical pulping. Third, the pulp yield from mechanical processing is roughly twice as much as the yield from chemical processing; the larger energy consumption of mechanical processing is compensated by smaller raw wood material consumption. Fourth, currently chemically produced pulp has more permanent optical and strength properties than mechanically produced pulp. Therefore, chemical pulp can be used for more purposes; mechanical pulp is normally limited to products with a relatively short lifetime.

In chemical pulping there are two prevailing process principles: the sulfate principle and the sulfite principle. These differ from each other by the type of cooking chemicals used to extract lignin from cellulose fibers. Until the 1960s sulfite technology had a major market share in pulp production, but during the past 20 years it has decreased rapidly and today represents less than 15% of the overall chemical pulp production in Western Europe (OECD, 1989). In

Scandinavia, its share is not greater than 10% of the total chemical pulp. In Central Europe, the share is still about 18%. This process is considered more efficient in processing the main raw wood material in the region (spruce) than the sulfate process. Chemical fiber has a firm position as a raw material for printing- and writing-quality papers not only in Western Europe but also worldwide. One explanation for this position is that the majority of the pulp (about 80%) is bleached. Chlorine is the prevailing bleaching agent (Judt, 1991). On the other hand, advanced cooking methods and substituting bleaching agents have substantially reduced chlorine use during the past 10 years.

Mechanical pulping is more widely used in Scandinavia than in Central Europe, due to a larger share of newsprint production for which mechanical pulp is the main raw material. Basically, there are two processing principles for mechanical pulping: grinding and refining the wood. For both technologies there are several modifications concerning utilization of heat, pressure, chemicals, and so on. A major part of the mechanical pulp is also bleached. Instead of chlorine, however, different peroxides are frequently used as bleaching agents in modern mechanical pulping mills (CTS-Engineering, 1991).

About 80% of the total paper production in the European OECD countries fall into the following groups (OECD, 1989):

Newsprint	14%
Printing and writing	35%
Household and sanitary	6%
Liner board	10%
Fluting	9%
Folding boxboard	7%

Newsprint and printing and writing qualities are used for graphic products, as the other qualities are converted into different paper products like corrugated board, boxes, wrapping paper, toilet paper, and so on.

An estimated average furnish for different paper qualities in Western Europe in 1986 is given in *Table 4.1*. The largest share was chemical pulp, about 38%. Over 80% of this share was bleached, with about 65% being bleached sulfate pulp. Mechanical pulps were used in about 20% of the mixture and reuse pulp (recycled fibers)

Table 4.1. Average furnish profile of paper and board in 1,000 tonnes. The production numbers of different paper qualities are totals for Germany, France, Italy, the Netherlands, the UK, Austria, Sweden, and Finland in 1986.

Paper quality	Newsprint	Printing & writing	Liner board	Fluting	Folding boxboard	Household & sanitary	Total
Chemical pulp	559.4	7,692.4	2,304.5	553.1	1,510.6	638.8	13,258.9 (38%)
Mechanical pulp	4,010.8	2,895.1	0.0	0.0	368.5	302.1	7,576.4 (21%)
Semi-chemical pulp	0.0	0.0	0.0	758.6	0.0	0.0	758.6 (2%)
Reuse pulp	1,033.9	1,076.1	2,189.2	2,244.9	1,655.1	1,709.5	9,908.7 (28%)
Other material	0.0	3,887.9	0.0	0.0	0.0	0.0	3,887.9 (11%)
Total	5,604.1	15,551.4	4,493.7	3,556.7	3,534.2	2,650.3	35,390.4 (100%)

Source: Kyrklund, 1991; OECD, 1989.

in about 28% of the overall fiber furnish. Mechanical pulps were predominantly used for the production of newsprint and printing-and writing-paper qualities. Chemical pulps were most often used in folding boxboards and soft paper qualities. Semi-chemical pulp was mainly used for fluting production. Reuse pulp was mostly used for the components of corrugated board, liner, and fluting and for soft paper qualities. Semi-chemical pulp had only a marginal share, about 2% of the total furnish mix. The share of non-fibrous materials, clay, starch, coating materials, and so on, was estimated to be about 25% for printing and writing qualities, and for the other qualities it was estimated to be negligible. This corresponds to an overall average of about 10% for non-fibrous materials in base paper and board.

The furnish profile in Central Europe was somewhat different from the furnish profile in Scandinavia. Reuse pulp was clearly used more in Central European production. For liner board and fluting, reuse pulp had a share of about 70% in Central Europe, while in Scandinavia the share was only about 5%. Reuse pulp was used in about 25% of the newsprint in Central Europe, but was hardly used at all in Scandinavia. For soft paper qualities, reuse pulp was used almost equally in both areas, about 35%. Mechanical pulp was more widely used in Scandinavian production, and semi-chemical pulp was nearly only used in Scandinavia.

Reuse pulp is produced in two types of plants. For less-demanding paper qualities, the waste paper recycled is treated in mechanical re-pulping plants. The pulp is not chemically de-inked and is typically used for brown papers and boards, for bobbins, egg containers, and so on. For higher-quality products, the waste paper is re-pulped, chemically cleaned from pigments, and sometimes also bleached. De-inked recycled pulp is typically used for newsprint and soft papers.

Generally, mechanical re-pulping for low-quality papers constitutes the following steps:

1. Waste paper feed and pulping.
2. Pre-cleaning (mechanical).
3. Refining.
4. Final cleaning (mechanical).
5. Thickening and storing.
6. Drying and packaging (only for market pulp).

Pulping (1) is done with either batch or continuous pulpers. Caustic soda and aluminum sulfate are used for pH adjustment and dispersion agents, such as derivatives of glycol ether. Pale wires and larger plastic particles are removed with ragger. In pre-cleaning (2), solid impurities such as plastic flakes are extracted from the pulp with centrifuges and pressure sorters. In the refining step (3), pulp undergoes washing, sorting, and milling. After this step, refining pulp is again cleaned mechanically (4) and thickened (5) for storage. Market pulp is dried and packed for transportation, step (6).

Chemi-mechanical re-pulping for higher-quality products is a process where ink pigments and some other chemicals are separated from the paper and removed using either washing or foaming methods. In some cases the pulp is also bleached to meet the brightness demand of the paper manufacture. The mechanical treatment steps of the de-inked pulp are similar to the uncleaned pulp steps. Normally, the process constitutes the following steps:

1. Waste paper feed and pulping.
2. Pre-cleaning (mechanical).
3. Refining.
4. Pigment removal (chemi-mechanical).
5. Final cleaning (mechanical).
6. Thickening, bleaching (optional), and storing.
7. Drying and packaging (only for market pulp).

Mechanical reuse pulp tends to darken when treated in alkaline conditions. Darkening is prevented by adding highly oxidizing agents, such as hydrogen peroxide or sodium hypochlorite, in pulpers. Water glass is used as the main de-inking chemical. In addition, different chemicals are used depending on the method for removing pigments. During the bleaching step, pulp is heated and mixed with bleaching chemicals for from 20 minutes to 2 hours. Ultimately, to make the brightness of the fiber last longer, the pulp is stabilized with sulfuric acid.

The highest rates of waste paper utilization are often gained by means of selective recycling: low-quality papers and high-quality waste. For large-scale recycling, however, other cases of recycling also enter the equation; this means processing low-quality, highly printed, and mixed waste paper to produce high-quality paper, which would require heavy cleaning technologies for recycled

Table 4.2. Estimated fiber yields for different waste paper grades.

Waste paper grade	Estimated yield
Highly printed	88%
Lightly printed or unprinted	92%
Composite waste paper	80%
Mixed waste paper	88%

Source: CTS-Engineering, 1991.

fibers. Increased cleaning and refining would, in turn, imply less fiber yield from recycled waste paper. Repeated and continuous large-scale recycling would potentially also increase the need for non-fibrous materials removal from waste paper, which also would reduce the average fiber yield. *Table 4.2* presents estimated yield factors for the waste paper grades selected for the feasibility study (CTS-Engineering, 1991). These estimated factors take into account the removal of non-fibrous materials from recycled fibers. Compared with yield figures gained from experience in the industry, the estimates presented in *Table 4.2* may be high. On average they represent a yield level of about 90%, for an estimated overall composition of recycled waste paper. It should be stressed, however, that the estimates in *Table 4.2* are preliminary estimates, which need to be checked for large-scale recycling in the planned full-scale study.

In addition to fiber, either primary or recycled, varying amounts of different additives are used for base paper and board manufacturing. Additives are used for all paper qualities. Currently their relative mass, however, is very small in comparison with fibers except for printing-quality papers, for which they are currently estimated to have a furnish share of about 25% (Kyrklund, 1991). As for volume, the important additives used are starch, calcium carbonate, and titanium dioxide. The concepts used in the ongoing discussion on paper recycling vary and are often different in meaning. In this feasibility study, the following nomenclature and concepts are adopted to distinguish between the main steps of the paper cycle. First, we have separated the production of base paper and board (bulk raw material) and the conversion (products usable for consumers). In general, the former is called *paper and board production (manufacture)* or *base stock production* and the latter *paper product manufacture* or *conversion*. The outputs are generally called *base paper and board* or *base stock* and *paper products*, respectively. Base paper and board is the bulk raw material for conversion. Regionally, conversion is supplied by the producers' base stock production and imports. Regional base stock production, in turn, covers both the needs for conversion industry and exports. Paper and board consumption takes the outputs from conversion and paper products as inputs.

Second, *waste paper* refers to the general, gross waste paper output from consumption and manufacturing processes. *Waste paper fiber potential* or *reuse fiber potential* means the part of the waste paper which could be collected for recycling as raw material for base paper production (i.e., gross output excluding long-term products and sanitary papers disposed with sewage). Third, *recycled fiber* or *reuse fiber* is the net fiber output from re-pulping to be used in the base stock furnish. The composition of recycled fiber depends on the re-pulping technologies (i.e., removal of non-fibrous materials). Re-pulping takes waste paper (raw material) as input. Fiber balance is used as a general name for the mass balance of the paper cycle.

Waste paper as part of the base paper and board production (1% to 5%, CTS-Engineering, 1991) is recycled within the paper mill and, therefore, is not part of the external waste management and recycling process. This category of waste paper originates in different trimming and finishing steps in the base paper and board production lines. It is re-pulped and reused as feed stock for paper and board machines – often for the same machine it originally came from. This is normal practice and has no effect on the fiber balance of the recycling.

Waste paper produced by the paper product manufacture (conversion) is generally recycled either to base paper and board manufacture or to manufacture of special paper products like bobbins. Transportation is needed, however, since the product manufacture and the paper stock manufacture are normally in different locations. A large portion of the waste is lightly printed or unprinted. But also highly printed and composite wastes are generated. The amount of this type of waste is, however, relatively small – only 5% to 6% of the total waste amount (Schwab, 1991). The waste stock is often separated into fairly homogeneous grades by the manufacturers; therefore, the waste material is easy to use for fiber recovery.

Graphic products, newspapers, books, advertisement leaflets, and copy papers are distributed as such, but most paper products (beverage containers, wrapping papers, corrugated board boxes) are distributed with the actual primary product as its package. Packaging and distribution, however, usually produce negligible amounts of waste paper assuming that all the unpacking of the actual products, which produces a significant part of the waste paper currently collected for recycling, is included in the consumption segment of the cycle. In distribution, the weight of the product is the most important factor. On average, the specific weight of the package by volume of the primary product is relatively small.

In the case of light goods and small units (for example, some household chemicals and food items), the weight of the packaging can be from 10% to 15% of the total weight, and as such a relative change in the transportation demands caused by heavier packaging would not be negligible. If the recycling of fibers increases weight in respect to the primary volume of consumption (that is, number of newspapers and books per tonne of food, and so on), then the transportation energy inputs in the distribution segment also increase (that is, more logistical inputs).

For the most part, waste paper derives from the end-use of products. Part of the products end up in long-term use (in archives and on bookshelves). The share of these long-term products is estimated at 10% to 15% (Ebeling, 1991) in Western Europe. Household and sanitary papers cannot be recycled because of their use and the method of disposal. Toilet papers, which represent about 50% of this group and 5% of the total waste paper, are disposed as sewage. Other soft paper waste goes to municipal incinerators and landfills. The rest, about 75% to 80% of all paper and board, is produced in different grades and compositions for different waste-management systems. In this feasibility study on paper recycling in Western Europe, a four-grade classification system has been used to characterize paper waste:

1. Highly printed waste paper with a majority of newsprint.
2. Slightly printed or unprinted waste paper.
3. Composite waste papers and boards, e.g., liquid packages.
4. Mixed waste paper, a mixture of (1), (2), and (3).

Paper containing waste (municipal waste) comprising the paper disposed but not separated among general household waste might be considered a grade as well. This grade is not a technically or

Table 4.3. Estimated shares of different waste
paper grades from consumption in 1986.

Waste paper grade	Estimated share
Highly printed	40%
Lightly printed or unprinted	55%
Composite papers and boards	5%

economically feasible potential source of fibers. Even though it currently represents a relatively large part, about 60%, of the total waste paper fiber potential, it is not included in this feasibility study for the reasons mentioned above.

The fiber type of the waste paper is an important technical constraint of recycling since the properties of mechanical and chemical fibers imply different demands for paper machines. It was not possible to take this fact into account in the waste paper classification because the resources necessary to study the impact of the fiber type distinction were far beyond those available for this study.

An estimate of the relative amounts of different grades of waste paper in the Western European waste paper fiber potential is given in *Table 4.3*. This estimate is given for the waste paper generated in consumption before disposal; separation of grades has not been taken into account. The mixed grade (4) is introduced by the method of disposal and is determined by the level of separation. The estimate is based on the waste paper raw material quantities given in available waste paper statistics (OECD, 1989; FAO, 1990) and on an adjustment of the collection rates (*Table 4.4*) for meeting the reported total waste paper input for pulp production. Currently the share of highly printed waste (1) is estimated to be about 40% of the total waste paper fiber potential and it has been growing rapidly during the last two decades in Western Europe, especially in Scandinavian countries. One reason for the growth has been the rapid increase in multicolor advertisement leaflets used in direct advertising to households. The share of lightly printed or unprinted grade (2) is estimated at 55% and the share of composite waste papers and boards (3) is estimated to be 5% (Sandberg, 1990).

The reuse potential of waste paper fibers is highly dependent on the collection and separation rates of disposal. In *Table 4.4* an estimate is given of these factors. In addition, an estimated sewage rate is also provided.

Table 4.4. Estimated collection, separation, and sewage rates of consumption for 1986, in percent.

	Source of waste					
	News-	Printing &	Liner	Flut-	Folding	Household
	print	writing	board	ing	boxboard	& sanitary
Collection rate	40	35	50	50	30	0
Separation rate	80	0	30	30	20	0
Sewage rate	0	0	0	0	0	50

The sewage rate, which applies to household waste only, is not part of the reuse fiber potential. What is left after sewage is either collected for fiber recycling or dumped into the municipal waste system. The relative share allotted to separation indicates the amount of the collected waste separated into homogeneous groups based on their characteristics. The rest of the collected waste remains mixed (4). Based on the rates given in *Table 4.4*, a rough estimate can be derived for the final destination of the base paper and board consumed in Western Europe (*Table 4.5*): long-term products about 10%, municipal waste about 50%, sewage about 5%, and utilized reuse fiber potential about 35%. In *Table 4.5*, the estimated amount of waste paper coming from the conversion is 5%. The theoretical total reuse fiber potential is about 75%.

The utilized reuse fiber potential, currently approximately 35% of all base paper and board consumption, is estimated to contain the following grades: highly printed waste paper (1) about 20%, lightly printed or unprinted waste paper (2) about 17%, composite waste paper (3) about 1%, and mixed waste paper (4) about 60% (see *Table 4.6*). These figures also include the estimated wastes from product manufacture.

To meet the physical property demands of the use of the paper and board products and their manufacture, there are a number of ways of modifying and furnishing the base paper or board. As discussed earlier, recycled fibers lose some of the properties which primary fibers have. For example, re-pulping causes fibers to shorten, which, in turn, reduces strength and moisture properties.

Continuous increased large-scale recycling can also introduce fiber-aging problems with changed distributions of the ages and a higher proportion of older fibers. At present, this is not a problem due to a relatively low overall recycling rate; it is possible to

Table 4.5. Estimates of production and final destination of selected paper and board qualities in 1,000 tonnes. Estimated waste from conversion (5%) is included. The production numbers are totals for Germany, France, Italy, the Netherlands, the UK, Austria, Sweden, and Finland in 1986.

Paper quality	Newsprint	Printing & writing	Liner board	Fluting	Folding boxboard	Household & sanitary	Total
Production	5,604.1	15,551.4	4,493.7	3,556.7	2,967.6	2,650.3	34,823.8 (100%)
Output as:							
Long-term consumption	0.0	3,308.3	0.0	0.0	0.0	0.0	3,308.3 (10%)
Municipal waste	3,194.3	7,452.6	2,134.5	1,689.4	1,973.5	1,258.9	17,703.2 (51%)
Sewage	0.0	0.0	0.0	0.0	0.0	1,258.9	1,258.9 (4%)
Utilized reuse fiber potential	2,409.8	4,790.5	2,359.2	1,867.3	994.1	132.5	12,553.4 (36%)

Table 4.6. Estimated grade profile of available waste paper in 1,000 tonnes. Estimates are based on the shares given in *Table 4.3* and disposal characteristics given in *Table 4.4*. Estimated waste from the product manufacture (5%) is included. The base production numbers of different paper qualities are totals for Germany, France, Italy, the Netherlands, the UK, Austria, Sweden, and Finland in 1986.

Paper quality	Newsprint	Printing & writing	Liner board	Fluting	Folding boxboard	Household & sanitary	Total
Highly printed waste paper	1,787.7	233.3	195.5	154.7	39.2	39.8	2,450.1 (20%)
Lightly printed waste paper	196.1	544.3	669.6	529.9	119.6	92.8	2,152.3 (17%)
Composite waste paper	0.0	0.0	0.0	0.0	158.8	0.0	158.8 (1%)
Mixed waste paper	425.9	4,012.9	1,494.2	1,182.6	676.6	0.0	7,792.2 (62%)
Total	2,409.8	4,790.5	2,359.2	1,867.3	994.1	132.5	12,553.4 (100%)

fulfill the demand for recycled waste paper with such papers that are, in practice, entirely based on primary fiber (newsprint, printing and writing, primary liner board, and so on). There is no need for or even the possibility of collecting papers containing recycled fiber such as sanitary papers and low-grade shoe boxes for recycling. Consequently, today's recycling is predominantly based on reusing paper fibers only once. *Figure 4.2* illustrates a theoretical age distribution of fibers in waste paper. The figure gives theoretical distributions of fiber age for stationary and continuing recycling in a closed system with the furnish share of recycled fibers as a dependent parameter. No separation of undersized fibers nor fiber losses in recovery have been taken into account. Assuming, for example, that 70% of recycled fiber is used for all papers and boards continuously, in the long run, more than 20% of the fiber employed for recovery would be used more than 5 times and about 5% more than 10 times. However, depending on the paper grade, a fiber can be reused only 3 to 5 times before it fragments or dissolves. This means that all of the fiber collected for recycling cannot be utilized in a closed, stationary recycling system. Due to technical and physical limitations the overaged part of fibers needs to be extracted in re-pulping to maintain the continuity of paper production.

In addition, in recovery there are fiber losses in the usable percentage of fibers, which correspond to about 5%. Considering these practical limitations, a theoretical maximum share for reuse fibers in the overall furnish, with a 100% collection rate for all waste paper, can be estimated at 75% to 80% depending on the maximum number of the reuse cycles accepted (see *Figure 4.3*).

However, the maximum collection rate cannot reach 100% because of long-term products, household papers, and paper exports. Excluding the part which is not recyclable and assuming a 100% collection rate for the rest, theoretically a 65% to 70% overall furnish share could be reached. In practice, the collection rate is never that high. The highest level that has been achieved so far is 55% in the Netherlands (Kyrklund, 1991). Under these conditions, a furnish share of about 50% seems achievable as an average for Western Europe. However, the utilization of waste paper fibers in some European countries has risen to 65%. This can be explained by a consumption-dominated paper balance and by waste paper imports.

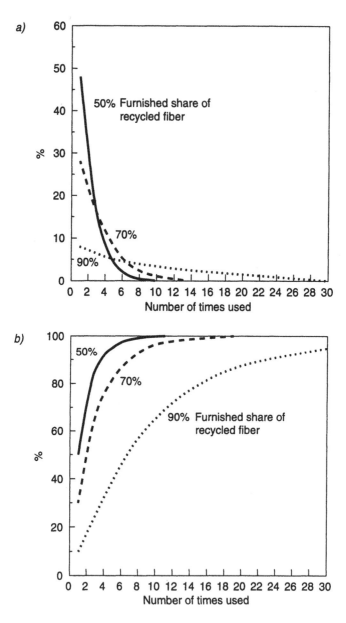

Figure 4.2. Theoretical age distribution of fibers in waste paper for continuous recycling in a closed system with the share of furnish as a parameter: (a) distribution and (b) cumulative (integral of a). The furnish share is assumed to be constant over all paper qualities.

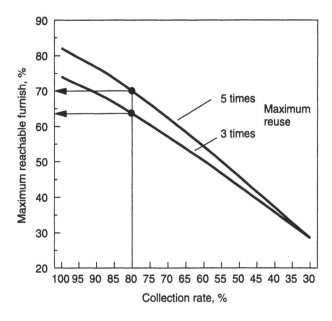

Figure 4.3. Maximum reachable share of reuse fibers in furnish for continuous recycling in a closed system with accepted number of cycles as a parameter (theoretical).

Depending on the average age of the recycled fibers used in the furnish, the thickness of the base paper or board should be increased to maintain the required strength and scoring level. Alternatively, undersized fibers could be screened out of the stock; however, because other fiber properties are also involved, such as strength, scoring, and folding, screening alone would not necessarily guarantee the adequate properties. It would also mean a smaller yield of fibers because the oldest part is always screened out or dissolved. The normal solution to achieve the properties required is to add primary fibers to the furnish either by mixing them with reuse fibers or by assembling the paper as layers of different quality fibers. The specific weight of papers in the cases where recycled fiber is used tends to be 5% to 15% higher than in the cases where only primary fibers are used (Judt, 1991).

Chapter 5

Model Treatment of the Paper-Recycling System

A generic process network model called International Database for Ecoprofile Analysis (IDEA) developed at IIASA (Lübkert *et al.*, 1991) was used to establish the inventory, employing the unit process database for generating production structures and for the inventory calculations starting from a given final product demand. The unit process database contains the data of the inputs and outputs of the individual processes and the data for the production and the post-processing profiles. The database defines the production shares in the cases where there are two or more processes producing (post-processing) the same product. Substitution data for alternative products (namely, orientation profiles between recycling and waste management) and transportation profiles are also included.

Currently, the IDEA database contains approximately 400 unit processes and about 400 products. The data represent the state of the art in the late 1980s in production, transport, and so on. The unit processes describe either entirely or in part the following 11 economic sectors:

1. Fuel, minerals, and mining.
2. Energy conversion.
3. Goods transportation.
4. Inorganic chemicals production.
5. Organic chemicals production.
6. Glass production.

   7. Pulp and paper production.
   8. Aluminum production.
   9. Iron and steel production.
 10. Beverage container manufacture.
 11. Waste management and recycling.

For this feasibility study on paper recycling in Western Europe the IDEA database was supplemented with data on the specific area of paper recycling. However, these data, comprising transportation and processing waste paper for reuse, are preliminary and are intended to be expanded and verified in a planned full-scale study on paper recycling.

The solution method employed by IDEA is iterative and uses the effective relational information. The algorithm proceeds step-wise through the system, balancing the imbalanced subjects and adjusting the related objects. The imbalanced objects become subjects in the following step. The first set of subjects are those affected by the boundary demands. The iteration terminates when there are no imbalanced subjects left to be adjusted.

The inventory model (IDEA) is currently capable of solving static systems only. It cannot be used for solving any balances other than simple substitutions. For example, the availability of waste paper cannot be taken into account, since all the profiles of alternative production technologies must be given as constants. Therefore, the framework of the inventory for each recycling scenario with collection rates and so on has been created separately with a worksheet program. In this section the main principles and procedures of the worksheet model are briefly discussed.

The principle of the worksheet model is based on the paper-fiber balance of Western Europe. The model computes the waste streams, available reuse fiber potentials and demands, and waste paper exchange between two regions of production and consumption. In this case the regions are Scandinavia, represented by Sweden and Finland, and Central Europe, represented by Germany, France, Italy, the Netherlands, the UK, and Austria. The method of computation makes use of the profiles for product characteristics, waste paper collection, and other factors given to it as parameters. The basic solution procedure, which is the same for both regions, begins at the paper supply and follows the principal flowchart in *Figure 4.1*. The steps involved are:

1. Paper supply.
2. Conversion (paper product manufacture).
3. Consumption.
4. Consumer disposal.
5. Waste paper collection.
6. Paper and board production.
7. Waste paper raw material balance.

For the fiber balance solution, the paper volume is broken down to seven quality groups following the commonly adopted statistical practice (OECD). These quality groups are the following:

1. Newsprint qualities.
2. Printing and writing qualities.
3. Liner board.
4. Fluting (corrugating medium).
5. Folding boxboard qualities.
6. Household and sanitary qualities.
7. Other qualities.

First, for each quality a supply balance based on the statistics is given in terms of import and production and, on the other hand, export and consumption. At this stage, consumption is the total input to graphic product and paper product manufacture. The balance equation for the supply of each paper quality is simply:

$$import = (consumption + export) - production. \qquad (5.1)$$

The balance of product manufacture is calculated next using the waste profile given as a table of parameters defining the estimated total losses and their distribution in different waste paper grades (see *Table 5.1*). The profile is given separately for graphic products and other paper products. The figures given in *Table 5.1* are, at this stage, only rough estimates of the output profile of the grades. On the other hand, the profiles of grades have not been directly taken into account in the inventory calculations due to the simplified presentation of re-pulping for the unit processes. In fact, reuse fiber production has been represented by only one unit process estimated to represent an average for all reuse fiber production from all waste paper grades. Accordingly, the inventory accomplished in this feasibility study does not handle different waste paper grades separately but treats the grades as one aggregated, average grade. The worksheet model is, however, designed for solving a more complete fiber balance intended to be accomplished in the planned full-scale study. Therefore, the breakdown of waste paper to different grades has, in principle, been included in the concept.

All the waste paper stemming from manufacturing is assumed to be sorted and collected for reuse. In practice, part of the waste paper is mixed and all waste paper is not collected for reuse as paper raw material. Due to its relatively minor importance and lack of data, this assumption is appropriate in this version of a balance model. The same profile data are used for both Central Europe and Scandinavia.

The balance equations for product manufacturing for each paper quality are:

$$output \; = \; input \; - \; losses \tag{5.2a}$$

$$reuse\; fiber\; potential = losses. \tag{5.2b}$$

The distribution of losses in different waste paper grade groups (x) is calculated as:

$$W_{x,y} = \sum_i \xi_{x,y}^{(i)} W_y^{(i)} \; , \tag{5.2c}$$

where $W_{x,y}$ is the total amount of waste paper grade x from all manufacturing of paper quality y; i is an element of {graphic product manufacture, paper product manufacture} (manufacturing sectors); $\xi_{x,y}^{(i)}$ is the share of the waste paper grade group x in the waste paper from the processing of paper quality y in the manufacturing sector i; $W_y^{(i)}$ are the total losses for processing paper quality y in the manufacturing sector i; $W_y^{(i)}$ is $v_y^{(i)} I_y^{(i)}$ where $v_y^{(i)}$ are the total relative losses for the paper quality y in the manufacturing sector i and $I_y^{(i)}$ is the input of paper quality y into the manufacturing sector i.

The balance of product manufacturing has been simplified to include only the paper component. Additional materials of the paper products, like plastics for beverage containers, have been excluded from the balance and, accordingly, from every other balance in the model. This, of course, distorts the material streams for some waste paper grades and, in theory, for the aggregated grade of the inventory calculations. Because of the relatively small volume of such materials (composites), however, the quantitative error in the

Table 5.1. Estimated waste profile for the conversion (paper product manufacture), in percent.

	Newsprint	Printing & writing	Liner board	Fluting	Folding boxboard	Household & sanitary	Other
Graphic products							
Losses of which	5.0	5.0	5.0	5.0	5.0	5.0	1.0
Highly printed	30.0	30.0	30.0	30.0	30.0	30.0	30.0
Lightly printed	70.0	70.0	70.0	70.0	70.0	70.0	70.0
Composite	0.0	0.0	0.0	0.0	0.0	0.0	0.0
Other paper products							
Losses of which	5.0	5.0	5.0	5.0	5.0	5.0	1.0
Highly printed	30.0	30.0	30.0	30.0	22.5	30.0	30.0
Lightly printed	70.0	70.0	70.0	70.0	52.5	70.0	70.0
Composite	0.0	0.0	0.0	0.0	25.0	0.0	0.0

Table 5.2. The paper product characteristic profile, in percent.

	Newsprint	Printing & writing	Liner board	Fluting	Folding boxboard	Household & sanitary	Other	Waste paper import
Long-term consumption	0.0	24.0	0.0	0.0	0.0	0.0	17.5	0.0
Highly printed waste paper	100.0	30.0	20.0	20.0	10.0	10.0	16.5	20.0
Lightly printed waste paper	0.0	46.0	80.0	80.0	40.0	90.0	66.0	80.0
Composite waste paper	0.0	0.0	0.0	0.0	50.0	0.0	0.0	0.0

overall streams can be expected to be rather insignificant. This assumption will be reconsidered in the planned full-scale study.

The balance of actual consumption (end-use) constitutes the rates of long-term consumption and wastes. The end-use balance, together with the conversion losses, determines the final destination of the paper and board supplied to the conversion. At this stage the output is given in measures of the estimated total amounts without taking into account the ways of consumer disposal. Waste categories like mixed waste paper, municipal waste, and available waste paper potential enter the equation after the consumer disposal. Distribution of the waste output in different grades is calculated using a paper product characteristic profile (see *Table 5.2*).

The product characteristic profile defines the share of the long-term consumption and the three grades of waste paper coming from different paper qualities consumed. The grades and shares included in the paper product characteristic profile describe the status of the conversion for different input paper qualities. The profile does not take into account consumer behavior in disposing the waste from households. The same profile is used for both Scandinavia and Central Europe. The balance equation for each quality (y) is

$$I_y = \sum_x \xi_{x,y} I_y \;,\tag{5.3}$$

where I_y is the input of paper quality y into consumption; $\xi_{x,y}$ is the share of waste paper grade x from the consumed paper quality y; x is an element of {long-term consumption and the three waste paper grades}.

The step following consumption is the consumer disposal. This step is needed to establish the actual structure of waste production. Consumer dependent waste paper categories (mixed waste, sewage, municipal waste) are introduced as outputs from the consumer disposal in addition to the *pure* waste categories. The balance is calculated using parameter-defining rates for sewage, collection, and separation (see *Table 4.4*). These parameters are assumed to be specific for the original paper qualities. Collection and separation rates are assumed to be constant for the three pure waste groups originating in a certain paper quality.

The sewage statistics apply to toilet paper only. The rest of the household paper is either collected or disposed of as municipal waste. Part of the collected waste paper, defined by the separation rates, is sorted. The remaining part is considered a mixture of the *pure* waste paper grades.

Collection rates are the main parameters in the fiber balance, since they define almost completely the raw material potential for recycling. Their regional differences also imply the transportation pattern needed for providing the pulp mills with recycled material. The separation profile in part determines the level of technologies needed for fiber recycling. Low separation rate implies heavier cleaning technologies, and lighter cleaning technologies are required for a high rate of separation. Separation rates are taken into account in the fiber balance model, but, as mentioned earlier, they have been excluded from the impact inventory version of this feasibility study for the sake of simplicity.

The balance equations for consumer disposals are:

$$W_{sw,y} = v_y W'_y \tag{5.4a}$$

$$W_{mw,y} = (1 - \eta_y)(1 - v_y)W'_y \tag{5.4b}$$

$$W_{mix,y} = (1 - \xi_y)\eta_y(1 - v_y)W'_y \tag{5.4c}$$

$$W_{x,y} = \xi_y \eta_y (1 - v_y)W'_{x,y} \tag{5.4d}$$

$$W_{sw,y} + W_{mw,y} + W_{mix,y} + W_{x,y_x} = W'_y , \tag{5.4e}$$

where v_y is the sewage rate for waste paper originating with paper quality y; η_y is the collection rate for waste paper originating with paper quality y; ξ_y is the separation rate for waste paper originating with paper quality y; $W_{sw,y}$ is the amount of sewage waste paper originating with paper quality y; $W_{mw,y}$ is the amount of municipal waste originating with paper quality y; $W_{mix,y}$ is the amount of mixed waste paper (collected) originating with paper quality y; $W_{x,y}$ is the amount of *pure* waste paper grade x (collected) originating with paper quality y; W'_y is the total amount of waste paper from consumption originating with paper quality y; and $W'_{x,y}$ is the total amount of waste paper grade x from consumption originating with paper quality y.

Within the waste paper collection step, the waste paper streams from product manufacture and consumption are summed. The balance is simply

$$output = input. \tag{5.5}$$

The balance of waste paper raw material is built up from waste paper collection for supply and the paper and board production for demand.

The pulp needs for paper and board production are calculated based on a given furnish profile. The calculated reuse pulp demand is converted into waste paper raw material consumption using the average yield rate resulting from the shares of different waste paper grades and the respective yields of fiber from reuse pulping. Consumption is satisfied with the collection and import of waste paper. In case consumption is greater than collection, the difference is provided by imports. On the other hand, if consumption is smaller than collection, the remaining surplus is removed by waste paper export and by disposal. In the event that imports are needed in one region (for example, Scandinavia), the other region (Central Europe) is given the highest priority as the import source. If the import demand is smaller than or equal to the surplus of the other region, no other world import is needed and the remaining waste paper surplus in the import source is assumed to be accounted for by disposals. The waste paper balance solution is based on the following equations:

$$W_i^{(I)} = \max\{(1/\eta_i^- V_i - W_i^{(C)}), 0\} \tag{5.6a}$$

$$W_i^{(S)} = \max\{(W_i -^{(C)} 1/\eta_i^- V_i), 0\} \tag{5.6b}$$

$$W_{i,j}^{(I)} = \min\{W_j^{(S)}, W_i^{(I)}\}, i \neq j \tag{5.6c}$$

$$W_{i,o}^{(I)} = W_i^{(I)} - W_{i,j}^{(I)}, i \neq j , \tag{5.6d}$$

where $W_i^{(I)}$ is the waste paper import; $W_i^{(S)}$ is the waste paper surplus; $W_i^{(C)}$ is the waste paper collection; $W_{i,j}^{(I)}$ is the waste paper import from the other region; $W_{i,o}^{(I)}$ is the waste paper import from the rest of the world; V_i is the reuse fiber demand; η_i^- is the average fiber yield; and i, j is an element of {Central Europe, Scandinavia}. The relationship between the fiber balance variables is illustrated in *Figure 5.1*.

Due to the resources available for the feasibility study, the system has been simplified for the calculations in many ways. Also, the data used in the calculations for the essential manufacturing processes (pulping, paper manufacture, reuse pulp production) have

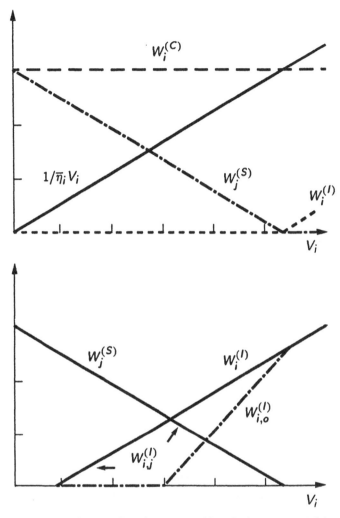

Figure 5.1. The relationship between fiber balance variables, equations (5.6a) to (5.6d).

been generalized and, to a large extent, are based on estimates. It proved impossible to collect data, at this stage, for a complete analysis. Accordingly, the basis of the impact calculations are sections of the total system investigated. Therefore, the results of the calculations only provide examples of the problem of the environmental impacts of recycling, and cannot be immediately applied to the overall system. The simplifications and generalizations employed in the current analysis are summarized as follows:

1. The boundary demand is set on the paper stock output (paper production) and on the disposal output (paper and board consumption). Accordingly, the steps found in between these borders (paper transport, conversion, distribution, and consumption, see *Figures 4.1* and *5.2*) are not included in the inventory. This simplification is based on the assumption that these steps are not affected, with regard to environmental impacts, by recycling. In reality, they are affected, for instance, because of the specific weight changes of the paper induced by the use of recycled fiber, which, in turn, implies changes in transportation demands. Both boundary conditions, however, are related to each other as intermediate states of a continuous cycle of paper fibers. These rates have been calculated with the separate balance program explained in Chapter 4.

2. The volume and pattern of paper and board production and primary consumption (numbers of products, their characteristics, and the geographical production and consumption patterns) are assumed to remain unchanged for all scenarios. The current production volumes are used for each country in Western Europe. No regional relocation of paper mills is assumed. Also the volumes of paper and board trade are assumed to remain constant.

3. No changes are assumed in area weight (grammage) or in any other specific weight of the paper. This, together with the preceding assumption, implies no changes in the total volumes (tonnages) of papers. Accordingly, there are no changes assumed in the transportation of papers, paper products, or waste paper due to the total volume change of transport requirements.

4. Generalized production data have been used for both primary and reuse fibers, namely, the quality of the pulp is assumed to be independent of the paper grade it is used for. The same production data are used for each pulp quality in both sectors. The furnish profiles, however, are sector specific. For primary fiber production there are four main pulp qualities: chemical (sulfate) unbleached, chemical (sulfate) bleached, chemical (sulfite) bleached, and thermo-mechanical bleached. Semi-chemical pulp has been allocated to unbleached chemical pulp. For reuse fiber production there are estimated data for one average pulp quality only, which is unbleached, de-inked pulp. All the pulp production is assumed to be integrated with paper

production, that is, no market pulp is included. The process data are presented in Appendix A.

5. Consequently, the pattern for pulp production is assumed to follow that of paper and board production. Pulp trade is omitted. This means that the pulp for paper and board in Scandinavia is assumed to be produced in Scandinavia. Pulp for Central European paper and board is assumed to be produced in Central Europe. This is valid also for waste paper pulp. This assumption, together with the assumption described in point 4, simplifies the present situation on the wood pulp market significantly. As stated earlier, about 30% of the wood pulp used in Central Europe was imported from outside the region in 1986. In addition, internal pulp trade between countries was about 5% of the total use in the region.

6. The profile of the primary pulp is assumed to be quality-wise constant for all scenarios. When primary pulp is substituted for waste paper pulp, the relative volumes of the different types of primary wood pulp in the total primary pulp are kept constant for each paper quality. This simplification overrules all potential, technical and economical factors implying changes in the primary pulp furnish for different rates of recycling. For example, price changes of different types of wood pulp due to the changes in the round wood market could be one factor affecting the overall consumption profile for primary wood pulp.

7. Most of the significant product streams have been traced back to raw materials. However, some products currently in the IDEA database do not have sufficient data for this purpose. For these products the inventory is incomplete.

8. CO_2 fixation in the biomass of growing wood has been taken into account in the inventory for the part employed by the wood pulp production. For the part which is, in theory, left unused in the forests (in the selective and maximum recycling scenarios but not in the zero recycling scenario), the CO_2 balance is assumed to be zero. Consequently, CH_4 originating in the decomposition of dead wood has not been taken into account.

9. CO_2 and CH_4 releases through the decomposition of harvesting wastes and waste paper deposited in landfills have been included in the life-cycle inventory by using roughly estimated unit processes. Time spans for biodegradation both in landfills and in forests are very long. Therefore, reliable estimations of

CH_4 and CO_2 releases are difficult to obtain. In this feasibility study estimates are based on so-called methane potential used for long-term methane releases. Methane potential is a measure of maximum methane yield from anaerobic biodegradation of organic compounds (Srivastava *et al.*, 1990), and therefore only a rough estimate of the actual release. The basis of the CH_4 release estimate is provided in *Table 5.3*. Because of the substantial difference in the heat absorption coefficients for CO_2 and CH_4, and consequently in their global-warming potentials, the release estimates are to be studied in-depth in the planned full-scale study. The sensitivity of the inventory results to the release estimates is discussed in Chapter 8. Paper exported out of Western Europe, long-term products, and sewage have been assumed to leave the system in the form of paper, which distorts the overall CO_2 and CH_4 balance to some extent. The boundaries of the trace back and the unit processes included in the production segment of the paper-recycling system can be seen in the data listed in Appendix A.

10. Some 80% of electric power for pulp and paper making is assumed to be produced on-site with natural gas as an auxiliary fuel. The remaining 20%, as well as the amount of electricity that is needed outside pulp and paper making, is estimated using an average European production profile. Energy conversion data used in the analysis are presented in Appendix B.

11. For all scenarios, 35% of the heat output of the incineration is assumed to be converted into electric power. The heat output (65%) is allocated to the heat demands of the production segment. The surplus left over after the allocation is not, however, further allocated. Thus, the potential credit from replacing fossil-fuel-based boiler heat has been omitted.

12. The transportation of waste paper between the regional sectors is assumed to be the same as for paper transportation. North America is assumed to be the source of imports from outside the regions studied. Transportation profiles used are listed in Appendix C.

Energy production and consumption are generally treated in gross measures in the inventory model. This means that for all unit processes, including primary pulp making, unit process data comprise energy (electric power, heat, and steam) demands in gross

Table 5.3. Approximation of the methane release from the decomposition of wood material.

Method of decomposition[a]		
Anaerobic digestion		
Ultimate analysis of dry matter (DM)[b]		
Carbon	40.0%	0.400 kg/kg,DM
Hydrogen	6.7%	0.067 kg/kg,DM
Other	53.3%	0.533 kg/kg,DM
CH_4 potential[c]		
3.976 kg CH_4/kg H		
Results		
CH_4 formation rate		0.267 kg/kg,DM
Carbon		0.200 kg/kg,DM
Hydrogen		0.067 kg/kg,DM

[a] Aerobic decomposition also takes place, but has not been taken into account in this study.
[b] Composition of wood material varies with species. Composition of cellulose (paper) also differs from that of wood biomass (harvesting wastes). Assumed composition is based on the mole shares of glucose ($C_6H_{12}O_6$).
[c] Rate of long-term methane formation; based on the assumption that hydrogen is completely transformed to methane.
Source: Srivastava *et al.*, 1990.

rates on the input side and the potential energy supplies on the output side.

Energy, like all the streams in the system, is treated according to the logical structure of the overall system comprising production, transportation, and energy conversion segments. The logical definitions for these segments are as follows:

1. *Production* segment comprises all the unit processes needed for the material supply and disposal network of the products studied. The leading principle is the dedication of unit processes. Processes belonging to the production segment are primarily operated for the purpose mentioned earlier. If there are by-products produced, like energy, they are taken into account in the overall balance, but they are still considered by-products.

2. *Transportation* segment comprises all the unit processes needed to provide the logistic output not only for the whole system studied, which is valid for the production and energy conversion segments, but also for its own need.

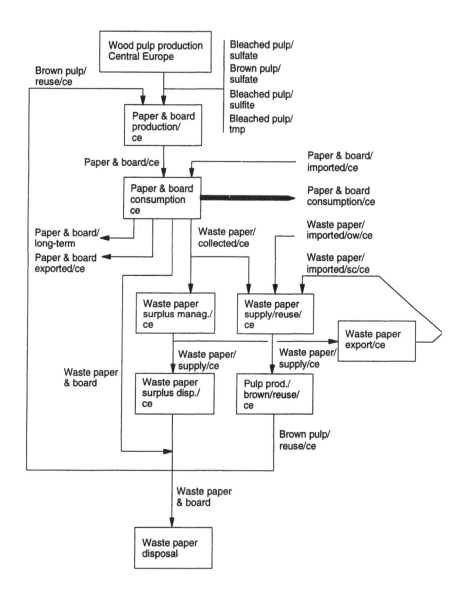

Figure 5.2. Flowchart of the waste paper recycling model.

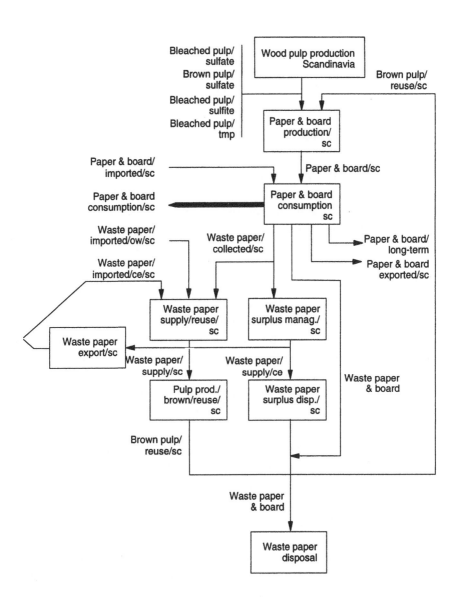

Figure 5.2. Continued.

3. *Energy conversion* segment comprises all the unit processes needed to provide the net energy demand for production and transportation segments and also for its own needs. Net energy demand is the difference of the total demands and supplies. According to this definition, the energy conversion segment comprises, besides the actual conversion processes, the unit processes for its material supply network. Thus, generally, production and energy conversion segments cover some of the same (basic production) processes.

According to this classification, for example, bark and soda boiler energy and energy from waste paper incineration are accounted for in the production segment. Additional energy produced in specific energy plants not based on industrial by-products but based on particular market fuels is considered part of the energy conversion segment. The principal scheme for modeling the paper-recycling system for impact calculations is presented in *Figure 5.2*. The consumption and production patterns have been illustrated for both Scandinavia and Central Europe. Waste paper is exchanged between these sectors with the exchange being based on waste paper surpluses, namely, the amount left over after the intra-sector recycling. The surpluses have the highest priority in fulfilling the import demands of the other region. If the surplus is insufficient, the rest is assumed to be imported from outside the regions.

The fiber balance for the IDEA model is primarily set up in both sectors with recycling rates (waste paper/collected/ce(sc)[1]) and furnish rates for primary fibers (bleached pulp/sulfate, brown pulp/sulfate, bleached pulp/sulfite, and bleached pulp/tmp[2]) and reuse fibers (brown pulp/reuse/ce(sc)) in paper production (paper and board production/ce(sc)). These rates are solved with a separate worksheet program for each of the three scenarios (M, S, Z). The reuse pulp production (pulp prod/brown/reuse/ce(sc)) is dedicated to fulfill the reuse fiber demand of paper production. Raw material supply for the reuse pulp production is represented by an auxiliary process (waste paper supply/reuse/ce(sc)) defining the rates for collected and imported waste paper (waste paper/imported/ce(sc)/sc(ce)) and (waste paper/imported/ow[3]/ce(sc)).

---

[1]ce(sc) denotes Central Europe (Scandinavia).

[2]tmp denotes thermomechanical pulp.

[3]ow denotes other wood.

These rates have been calculated with a separate balance program. The input of the collected waste paper to the reuse pulp production reduces the amount going to the waste paper surplus management (waste paper surplus manag./ce(sc)).

The import demand of waste paper is satisfied with waste paper export from the other sector and from outside the sectors studied. The shares of these two import sources are fixed on the basis of separate balance calculations. When defining the import rates, Western European waste paper (surplus) has been given the highest priority. Waste paper export (waste paper export/ce(sc)) of the surplus substitutes for (reduces) disposal (waste paper surplus disp/ce(sc)). Part of the waste paper, depending on the collection rate, goes directly to the disposal processes. For the disposal rate a fixed amount is used for incineration and landfill in the different scenarios. The characteristic profiles used for each scenario are listed in *Tables 5.4* to *5.6*. For the maximum recycling case (*Table 5.4*) a collection rate for the "dry" waste paper was assumed to be as high as 90% in both sectors. For soft papers and for the group labeled "Other," which mainly comprises grease-proof and parchment papers and boards, the collection rate of 50% percent was assumed. Based on these assumptions the overall, average furnish share for the reuse pulp was estimated at 56%. For paper disposal a profile of 26% was assumed for incineration and 74% was assumed for landfill, which corresponds to today's distribution in Western Europe.

For the selective recycling scenario (*Table 5.5*) the average collection rate was assumed to be raised to a level of about 47% of the total consumption from the current 34%. The estimated average furnish share for reuse pulp was 35% compared with the current 28%. In Scandinavia, the increased waste paper potential was assumed to be allocated to newsprint, fluting, and household and sanitary papers. In Central Europe, the share of the reuse fiber was assumed to be prominently increased for all papers except printing and writing papers. For paper disposal a profile of 26% and 74% was assumed, respectively, for incineration and landfill.

In the zero recycling scenario (*Table 5.6*) no recycling of paper for paper production was assumed. Instead, all waste paper from consumption was assumed to be incinerated (profile used 100% for incineration and 0% for landfill).

Table 5.4. Collection and furnish profiles for the maximum recycling scenario (M), in percent.

	Newsprint		Printing & writing		Liner board		Fluting		Folding boxboard		Household & sanitary		Other	
	Scandi-navia	Central Europe	Scandi-navia	Central Europe	Scandi-navia	Central Europe	Scandi-navia	Central Europe	Scandi-navia	Central Europe	Scandi-navia	Central Europe	Scandi-navia	Central Europe
Collection characteristics														
Collection	90	90	90	90	90	90	90	90	90	90	50	50	50	50
Separation	50	50	50	50	50	50	50	50	50	50	0	0	0	0
Sewage	0	0	0	0	0	0	0	0	0	0	50	50	0	0
Furnish profile for paper and board														
Chemical pulp	5	7	14	14	44	44	0	33	31	44	31	29	29	42
Bleached sulfate	90	80	80	80	0	0	0	0	80	66	90	80	90	80
Unbleached sulfate	0	0	0	0	100	85	100	85	20	16	0	0	0	0
Bleached sulfite	10	20	20	20	0	0	0	0	0	18	10	20	10	20
Unbleached sulfite	0	0	0	0	0	15	0	15	0	0	0	0	0	0
Mechanical pulp	39	37	5	5	0	0	0	0	13	0	13	15	15	2
Semi-chemical pulp	0	0	0	0	0	0	44	11	0	0	0	0	0	0
Reuse pulp	56	56	56	56	56	56	56	56	56	56	56	56	56	56
Other	0	0	25	25	0	0	0	0	0	0	0	0	0	0

Table 5.5. Collection and furnish profiles for the selective recycling scenario (S), in percent.

	Newsprint		Printing & writing		Liner board		Fluting		Folding boxboard		Household & sanitary		Other	
	Scandi-navia	Central Europe	Scandi-navia	Central Europe	Scandi-navia	Central Europe	Scandi-navia	Central Europe	Scandi-navia	Central Europe	Scandi-navia	Central Europe	Scandi-navia	Central Europe
Collection characteristics														
Collection	60	60	60	60	60	60	60	60	60	60	20	20	20	20
Separation	50	50	50	50	50	50	50	50	50	50	0	0	0	0
Sewage	0	0	0	0	0	0	0	0	0	0	50	50	0	0
Furnish profile for paper and board														
Chemical pulp	3	4	55	47	90	25	0	19	63	25	18	17	67	72
Bleached sulfate	90	80	80	80	0	0	0	0	80	66	90	80	90	80
Unbleached sulfate	0	0	0	0	100	85	100	85	20	16	0	0	0	0
Bleached sulfite	10	20	20	20	0	0	0	0	0	18	10	20	10	20
Unbleached sulfite	0	0	0	0	0	15	0	15	0	0	0	0	0	0
Mechanical pulp	22	21	20	18	0	0	0	0	27	0	7	8	33	3
Semi-chemical pulp	0	0	0	0	0	0	25	6	0	0	0	0	0	0
Reuse pulp	75	75	0	10	10	75	75	75	10	75	75	75	0	25
Other	0	0	25	25	0	0	0	0	0	0	0	0	0	0

Table 5.6. Collection and furnish profiles for the zero recycling scenario (Z), in percent.

	Newsprint		Printing & writing		Liner board		Fluting		Folding boxboard		Household & sanitary		Other	
	Scandi-navia	Central Europe	Scandi-navia	Central Europe	Scandi-navia	Central Europe	Scandi-navia	Central Europe	Scandi-navia	Central Europe	Scandi-navia	Central Europe	Scandi-navia	Central Europe
Collection characteristics														
Collection	0	0	0	0	0	0	0	0	0	0	0	0	0	0
Separation	0	0	0	0	0	0	0	0	0	0	0	0	0	0
Sewage	0	0	0	0	0	0	0	0	0	0	50	50	0	0
Furnish profile for paper and board														
Chemical pulp	11	15	55	54	10	100	0	76	70	10	70	67	67	96
Bleached sulfate	90	80	80	80	0	0	0	0	80	66	90	80	90	80
Unbleached sulfate	0	0	0	0	100	85	100	85	20	16	0	0	0	0
Bleached sulfite	10	20	20	20	0	0	0	0	0	18	10	20	10	20
Unbleached sulfite	0	0	0	0	0	15	0	15	0	0	0	0	0	0
Mechanical pulp	89	85	20	21	0	0	0	0	30	0	30	33	33	4
Semi-chemical pulp	0	0	0	0	0	0	10	24	0	0	0	0	0	0
Re-use pulp	0	0	0	0	0	0	0	0	0	0	0	0	0	0
Other	0	0	25	25	0	0	0	0	0	0	0	0	0	0

Chapter 6

Scenario Results

The scenario results in this chapter cover only Austria, Finland, France, Italy, the Netherlands, Sweden, the United Kingdom, and the former West Germany. Due to the preliminary nature of the data and the simplified approach, the inventory results must be interpreted with care; they show only possible trends of environmental impacts from recycling. In general, several of the environmental impact trends seem to be decreasing with increased recycling, but contraventional trends can also be seen. All the figures (*Figures 6.1 to 6.20*) are given for one kilotonne of paper and board consumed.

The trends for gross *electric power* consumption (*Figure 6.1*) and *heat* consumption (*Figure 6.2*) are linearly descending with increased recycling. The overall demand for electric power, as well as for heat and steam, is about 25% less in the maximum recycling case compared with the zero recycling case. For the selective recycling scenario the consumption is about 15% less than in the zero recycling scenario. The descending direction of the energy requirement reflects the differences in specific energy requirements between primary and reuse pulping. The electricity consumption and steam consumption for chemical pulp making, which is the main technology for primary pulp making, are relatively high compared with the re-pulping of waste paper. The additional energy consumption for primary pulp making is mostly covered by energy internally produced from burning black liquor and bark. The shift from wood-based energy in the production segment (or sector) to fossil-fuel-based production in the energy conversion segment is substantial with increased recycling. In the zero recycling case about 80% of

59

the total energy is based on wood, compared with only 45% in the maximum recycling scenario (*Figure 6.3*). The reduction trend in total energy consumption is most significant in the production sector. Energy conversion and transportation have only marginal roles in the overall energy demand.

The demand for *nonrenewable primary energy sources* (heat value potential, *Figure 6.4*) is about 100% larger in the maximum recycling case, while renewable resources are required about 60% less in this scenario in comparison with zero recycling (*Figure 6.5*). The reason for this is the substitution of bark and soda boiler steam and electric power, which are renewable energies, for coal, natural gas, and other nonrenewable fuel-based types of energy in the maximum recycling case. In the zero recycling case, on the other hand, most of the heat value potential of used paper is assumed to be utilized in the incinerators, which naturally lessens the demand for the actual energy conversion sector. In the maximum recycling case, a major part of the recoverable heat value of waste paper is assumed to be unused, that is, tombed with waste paper in landfills.

Because of increased demand for transportation and energy conversion with increased recycling, the overall *fossil fuel* (mass) inputs are about 75% greater in the maximum recycling case compared with the zero recycling case (*Figure 6.6*). Most of the growth in fossil-fuel use can be seen in an increase in the use of hard coal, brown coal, and diesel fuel for transportation (*Figure 6.7*). The prominent increase in coal fuels is due to the substitution of the residue wood-based energy production, which decreases more than the overall energy demand with recycling. Also, the assumption in the zero recycling scenario of a comprehensive incineration of waste paper has a substantial influence on the relative figures of increased use for coal fuels. The main factor which causes an increase in the transportation demand is the transportation of waste paper over considerable distances, for example, from Central Europe to Scandinavia. On the other hand, the reduction in log transportation counterbalances the overall transportation demand so that the total growth in energy required for transportation is rather insignificant.

Raw materials consumption prominently decreases with recycling (*Figure 6.8*). The amounts of raw materials (except wood) needed for wood pulp making (limestone rock, salt rock, and so on) are reduced by about 60%, following the ratio of the primary fiber quantities. It should be noted, however, that the calculations do not trace all material streams classified as raw materials because of the absence of data in the IDEA database. This distorts the figures for some specific raw materials. For example, the raw materials needed for de-inking agents are not included in the resulting figures. On the other hand, the predominant volume of raw materials consists of water and CO_2; any other material has only a marginal role in the overall raw material consumption.

For the main *air emissions* the maximum recycling case gives significantly more emissions of SO_2 (53%) and NO_x (7%) than the zero recycling scenario. On the other hand, emissions of other gases are significantly smaller in the maximum recycling scenario; CH_4 (–50%), CO (–30%), and gross CO_2 emissions (–45%).

The growth in SO_2 *emissions (Figure 6.9)* in the maximum recycling scenario is due to an increase in hard and brown coal combustion that replaces the combustion of wood-based fuels (bark, black liquor, and waste paper). The average sulfur content of these fuels is higher than that of wood-based materials. Sulfur emissions from primary pulp-making processes, other than combustion, are less significant.

Also, the increase in NO_x *emissions (Figure 6.10)* is due to the replacement of wood-based fuels. An increase in natural gas, as well as high-temperature combustion in general, is reflected in an increase in NO_x. Transportation has a noticeable role in NO_x emissions, but is less important in the general trend because there is no essential difference in total transportation requirements between different scenarios. This is perhaps surprising in view of the long distances involved in the assumed waste paper transportation from Central Europe to Scandinavia in the maximum recycling case. It should be noted, however, that the figures for transportation emissions include only direct emissions from vehicles. Thus, emissions given out during the production of electricity needed for electric trains are excluded from the transportation emissions' figures. A logical explanation of the trends in transportation requirements lies, also, in the ratio of the specific demand for logs to waste paper per final output. For a tonne of primary fiber, the mass of logs needed is roughly four times greater than the mass of waste paper required for reuse pulp. Thus, a gross CO_2 decrease in the volume of logs transported compensates for the longer distances over which waste paper must be transported.

The main factor explaining the differences in the CH_4 *emissions* (*Figure 6.11*) for different scenarios is the decomposition of harvesting wastes. Harvesting wastes represent about 35% of the total biomass of trees. This represents roughly the same amount of organic matter as the cellulose fibers in 65% of the biomass taken out as logs to the pulp mills. In each scenario, all harvesting wastes are assumed to be left in the forests where they decompose in five to ten years into CH_4 and CO_2. An alternative would be to collect the waste and burn it for energy production. Under the former assumption, decomposition of harvesting wastes becomes the main factor in CH_4 and CO_2 emissions – the more primary pulp required the greater the harvesting wastes and, eventually, decomposition gases. The approximate carbon balance for the production segment is provided in *Table 6.1*. When the "carbon turnover" increases with increased wood intake, both the amount of harvesting wastes left in the forest and the amount of the waste paper increase. Under the assumptions made, this implies increased methane potentials. It must be noted, however, that the specific methane release used is based on a theoretical maximum estimate. The sensitivity of the methane emissions to the release rate is discussed in Chapter 8. In addition, an interesting and important aspect related to this is how the life cycle of the forests influences emissions of the gases. This could not be studied in this feasibility study, but will be considered in the planned full-scale study.

The reduction of *gross CO_2 emissions* (*Figure 6.12*) depends on the transfer from carbon-rich (wood-based) fuels to less carbon-rich fuels (natural gas, hydropower, nuclear power) in the overall energy production of the total cradle-to-grave system of the scenarios. The same explanation can be given for the similar CO emission trend (*Figure 6.13*), even though harvesting with small, two-stroke chain saw engines makes a noticeable contribution to the amount of CO emissions.

However, when the formation of wood biomass (forest growth) is included in the calculations, the CO_2 trend is reversed. The calculated *net CO_2 emissions* (*Figure 6.14*) are negative for the production sector in all scenarios (meaning that a fixation of carbon takes place). In the zero recycling case the overall net emissions are −150% of the maximum recycling case. This can be explained by the carbon balance of paper production and consumption (see *Table 6.1*). Out of the total amount of carbon fixed in the wood biomass,

Table 6.1. Approximate carbon balance for the production segment.

	(M)aximum recycling scenario	(S)elective recycling scenario	(Z)ero recycling scenario
Carbon input as CO_2	531,366	855,990	1,341,523
Wood biomass formation	500,298	824,922	1,310,456
Potato biomass formation	31,067	31,067	31,067
Carbon output as CO_2	–314,759	–567,827	–1,045,387
Energy production at pulping	–184,348	–319,920	–487,325
Decomposition of harvesting wastes	–92,507	–152,433	–242,062
Waste paper incinceration	–15,896	–40,040	–315,999
Decomposition of waste paper	–22,008	–55,434	0
Carbon output as CH_4	–98,913	–178,081	–194,743
Decomposition of harvesting wastes	–74,442	–122,646	–194,743
Decomposition of waste paper	–24,471	–55,435	0
Carbon output as other			
carbon compounds	–117,694	–110,081	–101,393
Long-term consumption	–43,624	–43,624	–43,624
Net paper exports	–40,156	–40,156	–40,156
Sewage paper	–13,068	–13,068	–13,068
Mixed industrial & municipal waste	–20,846	–13,233	–4,545
Total carbon output	–531,366	–855,990	–1,341,523
Carbon transitions from input:			
CO_2	531,366	855,990	1,341,523
Carbon transitions to output:			
CO_2	314,759	567,827	1,045,387
CH_4	98,913	178,081	194,743
Other carbon compounds	117,694	110,081	101,393
Carbon compound balances[a]			
Input CO_2	1,948,341	3,138,628	4,918,918
Input CH_4	0	0	0
Input other carbon compounds	0	0	0
Output CO_2	–1,154,117	–2,082,033	–3,833,085
Output CH_4	–131,884	–237,442	–259,658
Output other carbon compounds	–294,234	–275,203	–253,482
Balance CO_2	794,224	1,056,595	1,085,832
Balance CH_4	–131,884	–237,442	–259,658
Balance other carbon compounds	–294,234	–275,203	–253,482

[a]Relative carbon masses: CO_2 = 27.3%; CH_4 = 75.0%; other = 40.0%.

35% is returned into the atmosphere through the decomposition of harvesting wastes. About 20% to 50% of the carbon decomposes into methane. Of the rest of the biomass traveling to pulp mills in the form of logs, roughly half ends up in paper. The other half is burned in bark and black liquor boilers to form carbon dioxide. About 20% of the carbon in paper ends up in archives or is exported out of Europe. About 25% of European waste paper is burned and the rest is deposited in landfill sites where eventually about 20% to 50% of it decomposes into CH_4 and the rest into CO_2. The reason why the zero recycling case appears to have less overall CO_2 emissions is obvious under the assumptions described above about the carbon balance. In the selective, and even the maximum, recycling case, the landfilled part of the waste paper releases, in a slow *burning* process, its heat potential and carbon into the environment. In the zero recycling scenario the burning takes place in boilers and the released heat is used as a substitute for heat produced, to a large extent, with fossil fuels. Thus the total CO_2 emissions are reduced by the amount corresponding to the substituted fossil fuels. In spite of the apparent simplicity in the current carbon balance, however, there are a number of unanswered questions left to be studied, such as the release rates of the methane and the carbon dioxide from the biodegradation of wood and waste paper, the paper streams leaving Europe, the role of archive or other long-term papers, and, eventually, the life-cycle of forests and the sustainable utilization of forest resources.

For *water emissions* a similar fluctuation in trends can be seen to that of air emissions. Total suspended solids (TSSs) are about 70% greater in the maximum recycling case than in the zero recycling case (*Figure 6.15*); biological oxygen demand (BOD) is about 10% greater in the maximum recycling scenario than in the zero recycling scenario (*Figure 6.16*). On the other hand, chemical oxygen demand (COD) and chlorinated organic compounds (AOX) emissions are significantly smaller (*Figures 6.17* and *6.18*). However, there are no acceptable explanations for all fluctuations; therefore one should be careful not to draw any further conclusions from this. The water emission data need to be updated to give more reliable results; for instance, the AOX emissions from reused pulp production should be investigated, as should the state and the performance of the wastewater purifiers.

The amount of gross municipal waste (the waste generated by consumption and manufacturing processes that is assumed to be further processed) decreases rapidly with increased recycling (*Figure 6.19*). However, if the terminal amounts (waste which is assumed not to be further processed) are considered the trend is reversed. The maximum recycling scenario produces about 50% less gross solid waste than the zero recycling scenario, although the amount of solid industrial waste is increased. The total net waste (terminal waste) in the maximum recycling scenario is twice that of the zero recycling case (*Figure 6.20*). The growth in industrial net waste volume with increased recycling is due to an increase in coal-based energy production, which implies increased coal-mining wastes, to which the ashes from the coal boilers also contribute.

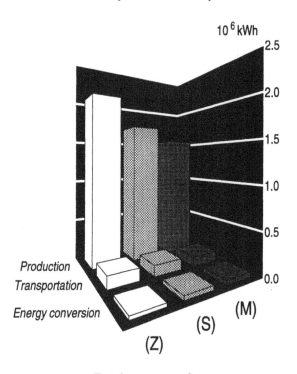

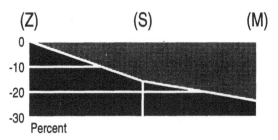

Figure 6.1. Trends in electric power consumption.

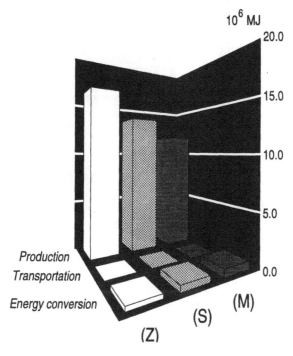

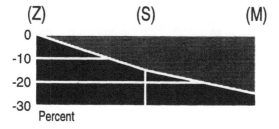

Figure 6.2. Trends in heat and steam consumption.

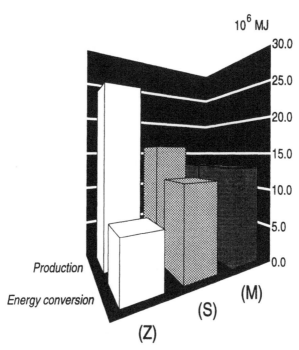

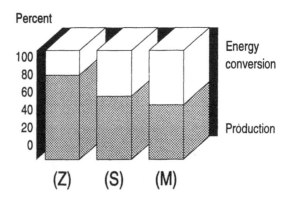

Figure 6.3. Distribution of energy production between production and energy conversion segments.

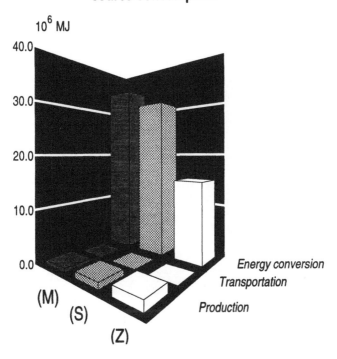

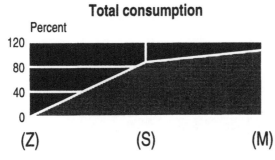

Figure 6.4. Trends in overall nonrenewable primary energy source consumption.

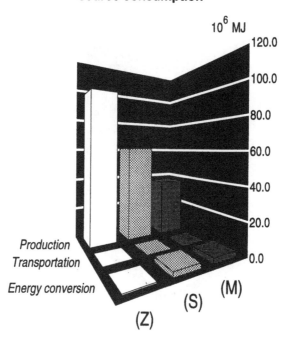

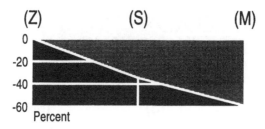

Figure 6.5. Trends in overall renewable primary energy source consumption.

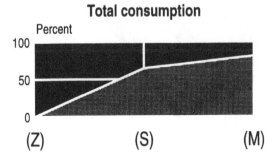

Figure 6.6. Trends in overall fossil fuel consumption.

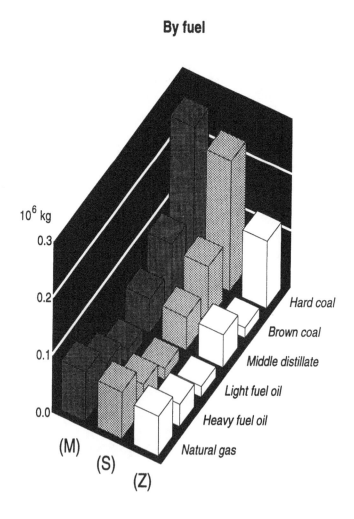

Figure 6.7. Trends in fossil fuel consumption, by fuel.

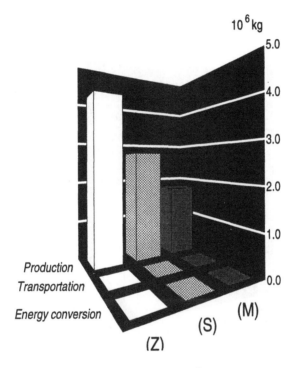

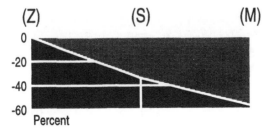

Figure 6.8. Trends in overall raw materials consumption.

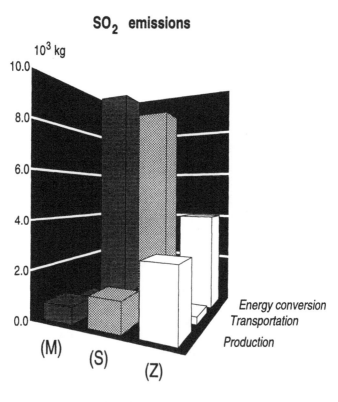

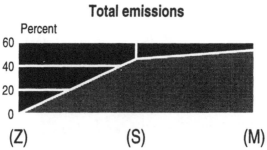

Figure 6.9. Trends in SO_2 emissions.

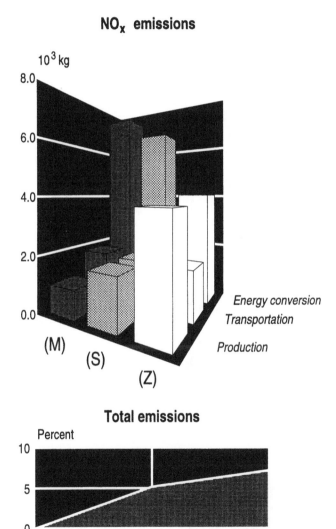

Figure 6.10. Trends in NO$_x$ emissions.

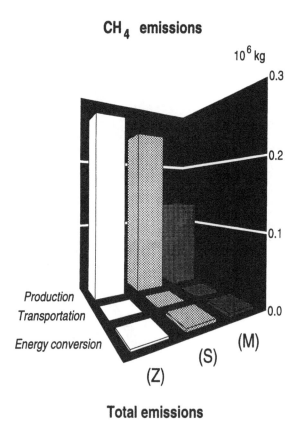

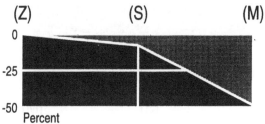

Figure 6.11. Trends in CH_4 emissions.

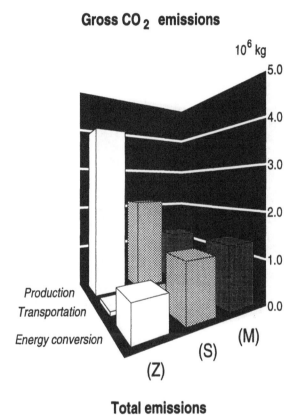

Gross CO$_2$ emissions

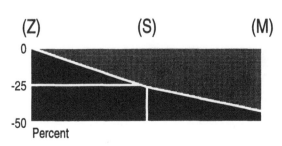

Figure 6.12. Trends in gross CO$_2$ emissions.

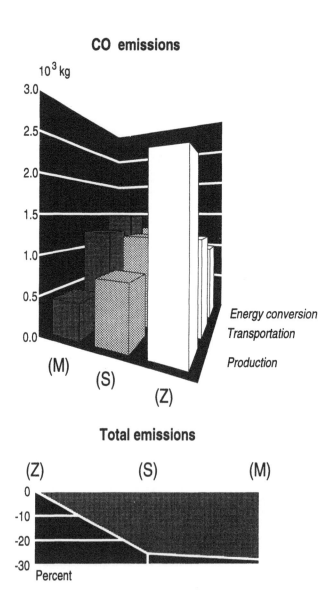

Figure 6.13. Trends in CO emissions.

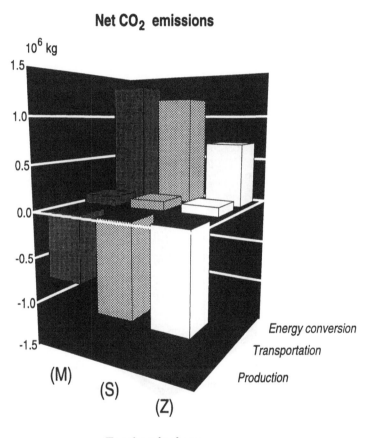

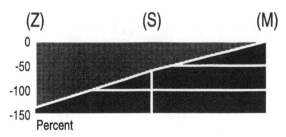

Figure 6.14. Trends in net CO_2 emissions.

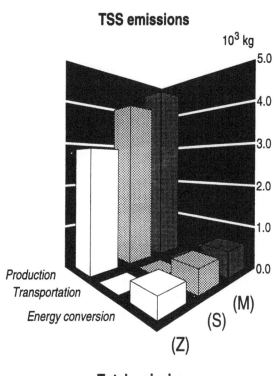

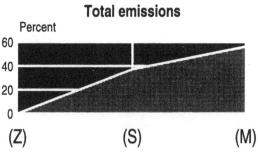

Figure 6.15. Trends in total suspended solids (TSSs).

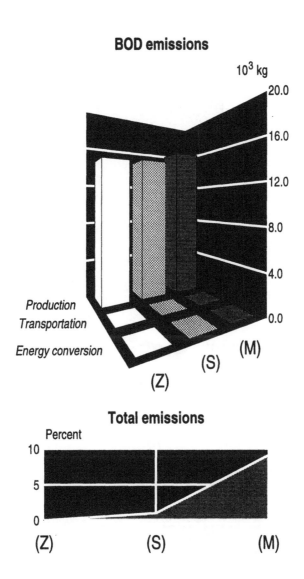

Figure 6.16. Trends in biological oxygen demand (BOD).

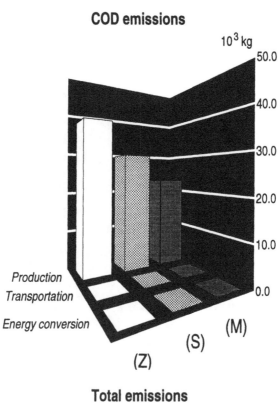

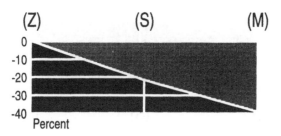

Figure 6.17. Trends in chemical oxygen demand (COD).

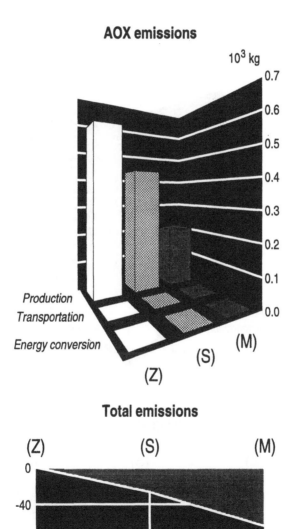

Figure 6.18. Trends in chlorinated organic compounds (AOXs) emissions.

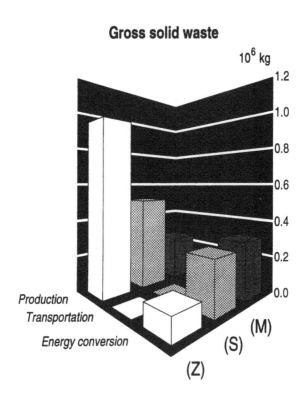

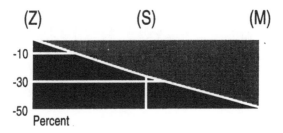

Figure 6.19. Trends in gross solid waste.

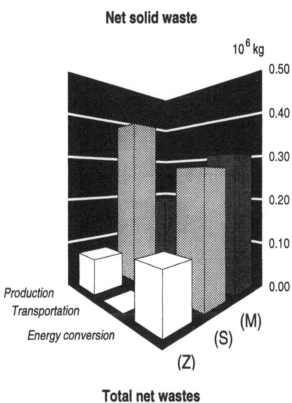

Net solid waste

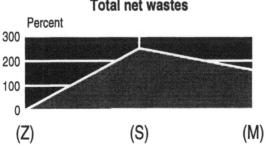

Total net wastes

Figure 6.20. Trends in net solid waste.

Chapter 7

Wood Balances

The scenario results concerning wood consumption for Central Europe and Scandinavia have been widened to encompass a wood balance for all of Western Europe (except Turkey and the former Yugoslavia). The wood balance is carried out for the year 1989, the most recent year for which complete statistics are available. Wood consumption is based on the production of final products in the region in 1989. To achieve this production in 1989, wood was imported to the region, but in reality wood was also exported from the region at the same time. In the actual wood balance, we have made a self-sufficiency calculation for the region, which means that the production of final products is assumed to be supplied entirely with wood from the region. Country-specific conversion factors for wood consumption have been employed for each forest product. These factors are from the mid-1980s (UN, 1986) and must be regarded as high estimates of wood consumption. In *Figure 7.1*, the wood balance is presented. The figure may require some clarification. Let us study the bar labeled "Production" in 1989 as a basis for this clarification. The wood consumption, expressed in million cubic meters solid over bark (s.o.b.), is presented for two major product aggregations (paper and board production and production of all other forest products). The required wood supply for the paper and board produced in Western Europe in 1989 was 116.3 million cubic meters s.o.b. The corresponding figure for the production of all other products was 228.4 million cubic meters s.o.b. In the production of wood for industrial manufacturing we get harvesting

Wood balance in Western Europe

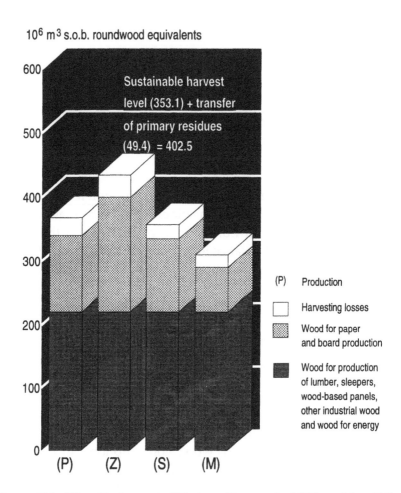

Figure 7.1. Wood balance in Western Europe in 1989 and for different paper-recycling scenarios.

losses of 23.4 million cubic meters s.o.b. In total, wood consumption would be 368.1 million cubic meters s.o.b. It should be pointed out that at the 1989 production level in Western Europe, raw material (wood and fibers) was both imported to and exported from the region. In the wood balance calculation we have made the assumption that the wood and fibers required for the actual production of industrial products in 1989 were produced domestically.

The wood balances for the different recycling scenarios are also presented in *Figure 7.1.* As expected, the selective recycling scenario has about 18% less total wood consumption than the zero recycling scenario. The corresponding figure for the maximum recycling scenario is about 27% less.

The long-term (next 100 years) sustainable biological harvest potential in the region (sustainable development of growing stock and flows from the stock), based on information from Nilsson *et al.* (1992), is estimated at 353.1 million cubic meters s.o.b. The transfer of primary residues (chips, sawdust, and so on) between industries should be added to the sustainable harvest level to get a complete wood balance. There are also residues from secondary processing available: residues from pulp mills and residues from recycled wood products (wooden packages, used furniture, and so on). These latter kinds of residues have been excluded from the actual wood balance. The transfer of primary residues constitutes 49.4 million cubic meters s.o.b. (stemming from the long-term sustainable harvest level). Thus, the total possible long-term sustainable wood supply is 402.5 million cubic meters s.o.b., which is indicated by the solid line across the figure. It should be pointed out that the estimated sustainable harvest levels presented by Nilsson *et al.* (1992, p. 163) are said to be conservative. It is also important to note that the region's actual roundwood production in 1989 was only 308.9 million cubic meters s.o.b. (FAO, 1991a), indicating that the forest resources were underutilized from the viewpoint of sustainability.

From the available results, it can be seen that wood consumption corresponding to production of final forest products in 1989 was well within the long-term sustainable limits of the harvest (about –35 million cubic meters s.o.b.). If there were no recycling, wood consumption would exceed the sustainable harvest level by about +33 million cubic meters s.o.b. With the increased recycling (selective and maximum recycling) scenarios wood consumption would be well within the sustainability limits (–45 and –87 million cubic meters s.o.b., respectively). From *Figure 7.1* it can also be seen that the results for wood consumption for "Production" in 1989 and the selective recycling scenario are quite similar. The selective scenario takes into account the geographical and economic availability of waste paper. It indicates that industry today is already close to the recycling rate estimated to be economically feasible and practical.

In carrying out the detailed calculations of the wood balance figure, it was established that no straight, linear relationship exists between increased paper recycling and decreased wood consumption. An increased recycling rate may result in increased usage of wood-consuming technologies (like chemical pulping instead of mechanical pulping) for the production of primary fibers. Such a shift depends on the availability and appropriateness of different waste paper qualities for paper production and the market conditions for different end-use products. We have not been able to analyze these effects to their full extent in this aggregated study. It will be a topic for further investigation in the planned full-scale study.

A change in the occurrence of recycling will have impacts on roundwood prices and silviculture and forest management. We have not been able to analyze the effects on roundwood prices in this study. Adams and Haynes (1991) have illustrated that an increase in the long-term recycling rate in the USA as a whole (from 29% to 39%) will decrease the long-term stumpage price in the southern states by some 20%. Wiseman (1990) has also pointed out that the stumpage price will be affected by an increased recycling rate. The increase in the recycling rate from selective recycling to maximum recycling in this study means an increase in average waste paper use from 39% to 56% in the pulp mixture. Therefore, at a greatly increased recycling rate, strong effects on the stumpage prices can be expected in Western Europe too.

One problem with increased recycling and decreased stumpage prices is the lower possibility of maintaining silvicultural standards and the vitality of the forest resources in Western Europe. To achieve long-term sustainable development of the forest resources and vital forests in Western Europe, Nilsson et al. (1992) have estimated that there would be an average yearly requirement of thinnings of roughly 121 million cubic meters s.o.b. in the region. About 80% of the raw material used in the late 1980s in Western Europe's pulp industry was in the form of pulp-logs (UN, 1991), of which the main part comes from thinnings. The remaining volume of fiber used in the pulp industry came from the residues of primary processing. By using this information, a rough balance for the thinning volume can be estimated. In *Figure 7.2*, the thinning volume required to

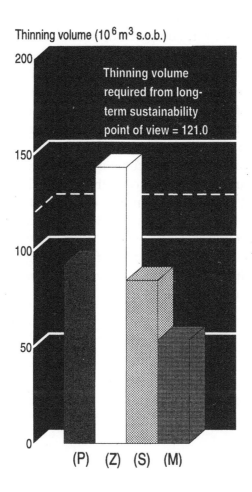

Pulp-log balance

Thinning volume (10^6 m^3 s.o.b.)

Thinning volume required from long-term sustainability point of view = 121.0

(P) (Z) (S) (M)

Figure 7.2. Pulp-log balance of paper and pulp production in Western Europe in 1989 and for different paper-recycling scenarios in comparison with the thinning volume required for long-term sustainability.

achieve sustainable development of the forest resources (according to Nilsson *et al.*, 1992) is compared with the pulp-log requirements for the 1989 production of paper and board and with the different recycling scenarios in Western Europe. It is assumed in these balances that the thinning volume is made up entirely of pulp-logs. In reality, some of the output from the thinnings is also in the form of saw-logs. From this comparison, it can be seen that, even at a domestic production level in Western Europe equivalent to the total number of pulp-logs required for the actual production of pulp and paper and board products in 1989, there is a gap between pulp-log consumption and the required thinning volume of 28 million cubic meters s.o.b. (due to unprofitable thinning operations). This gap will increase to 36 million cubic meters s.o.b. in the selective recycling scenario and 67 million cubic meters s.o.b. in the maximum recycling scenario. In the latter case, this means that more than 50% of the required thinnings may not be carried out. It can be argued that the thinning volume not used by the pulp industry can be used by the board industry. However, this is unlikely to occur because the board industry's paying capacity for pulp-log is much lower in comparison with the pulp industry, which will make the thinnings even more unprofitable.

Chapter 8

Sensitivity Considerations

Some of the trends identified in Chapter 7 are very sensitive to the data used and the assumptions made for the system studied. Assessing the quality of the results is, however, a major exercise which requires careful planning, substantial effort, and sufficient metadata (data on data) to assess the quality of the input data in the inventory model and the methodology (assumptions) used. Consequently, fulfilling all these requirements was beyond the possibilities of this feasibility study. However, to illustrate the importance of the quality assessment, in this chapter three examples are provided on the sensitivity of the scenario results to some key data and assumptions. The ranges of the data variations are arbitrarily chosen because the available metadata was not sufficient for assessing true fluctuations.

Example 1:   Emissions from the decomposition of
             harvesting wastes and waste paper.
Original data: 50% of the carbon released as CH_4.
             50% of the carbon released as CO_2.
Changed to: 20% of the carbon released as CH_4.
             80% of the carbon released as CO_2.

This change of release data affects the trends of the CH_4 and CO_2 emissions as indicated in *Figure 8.1*. The fluctuations are substantial for CH_4 and net CO_2 emissions (gross emissions minus fixation). For gross CO_2, the change is relatively small, but still about three times that of CH_4. Although the main directions of the trends seem to remain the same, the assessment of the global-warming effects is highly dependent on the mutual shares of the two greenhouse gases

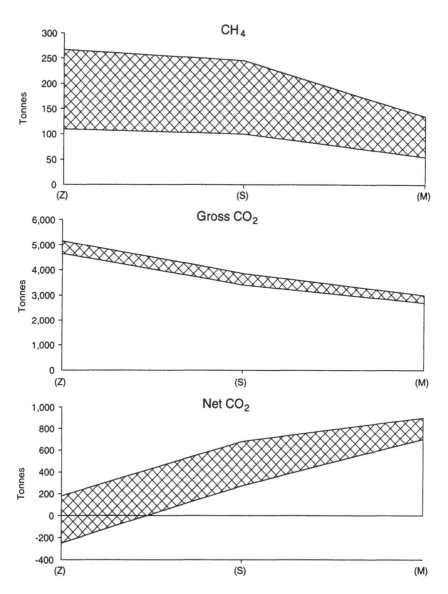

Figure 8.1. Effect of the decomposition data on methane, gross carbon dioxide, and net carbon dioxide emissions.

(CH_4 and net CO_2), which change substantially with the release data.

Example 2:   Natural cycle of the forests.
Original data: Not included.
Changed to: Include the assumed decomposition of the
                 unused wood biomass volume relative to the
                 zero recycling case.

Including or excluding the natural cycle of the forest strongly affects the trends of the greenhouse-gas emissions. The trends of both CH_4 and net CO_2 may even be reversed when the natural cycle is included (*Figure 8.2*). This indicates very well the importance of the boundaries set for the problem studied. Often, these are many times more significant than the actual data used, and should, therefore, be very carefully considered, justified, and reported. The example also points out the usefulness and necessity of the sensitivity analysis of the assumptions made on the system studied.

Example 3:   Bleaching of the reuse pulp.
Original data: All the volume unbleached.
Changed to: All the volume bleached using hydrogen
                 peroxide and 50% more electric power and
                 100% more steam.

Changing the assumption on the bleaching of the waste paper pulp affects more or less all trends due to the energy and intermediate product inputs (H_2O_2, NaOH) needed for bleaching. All trend changes are, however, much smaller than those indicated above. Some examples are provided in *Figure 8.3*. This illustrates the minor quantitative importance that marginal material and energy changes have in the overall system. In paper production, the main requirements for energy and materials come from paper machines and primary pulp production in all scenarios. Even in the maximum recycling case, less than 20% of the overall steam and electric power consumption is due to waste paper pulp production. Consequently, the response of even a relatively large change in its energy and material data is markably reduced on the overall system level.

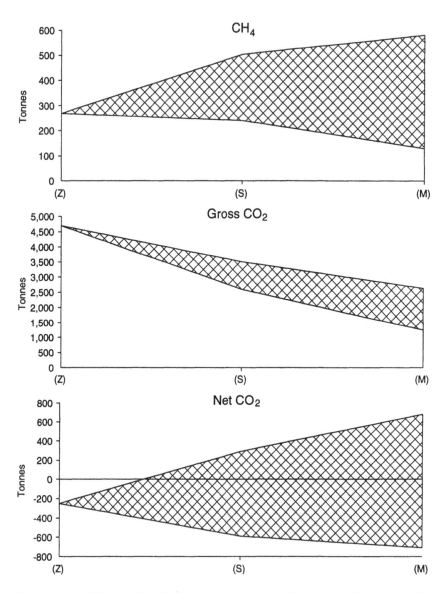

Figure 8.2. Effect of including or excluding the natural forest cycle on methane, gross carbon dioxide, and net carbon dioxide emissions.

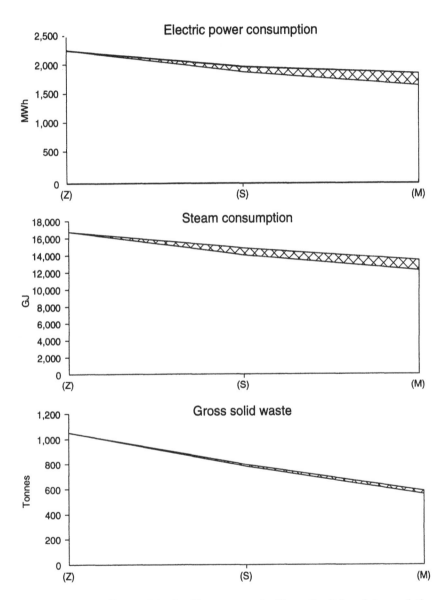

Figure 8.3. Effect of including or excluding the bleaching of the waste paper pulp on electric power, steam consumption, and gross solid wastes.

The role of the marginal operations should not, however, be underestimated in a life-cycle assessment. Even though each is a minor factor, their total contribution to material and energy demands and emissions may be significant. The results obtained in this feasibility study indicate about a 15% energy share for the processes feeding the actual paper and pulp production processes. Marginal operations are also important due to the hazardous substances that might be involved.

Chapter 9

Policy Implications

Judging the environmental friendliness of products using simple arguments such as the rate of recycled fiber may lead to errors of judgment. The reasons for this are many; most of them, however, relate to the generic nature of the overall environmental impacts of products. There are relations between different economic sectors involved whose implications cannot be directly foreseen because of their complexity. For example, reducing primary fiber consumption in paper production implies a reduced energy production from bark, lignin, and waste paper. However, since the overall energy demand is not reduced by the same proportion, additional energy production is needed. In our example, the current average Western European mixture of fuels was used for this purpose. The consequences can be seen as an increase in SO_2 emissions, and also a significant increase in the consumption of nonrenewable energy sources.

The results of the static calculations under the prevailing conditions in the late 1980s and under conditions of increased recycling rates are summarized in *Table 9.1*. It can be seen that an increased rate of recycling has a positive impact on some environmental parameters and a negative impact on others. The overall task for policy makers is to weigh the pros and cons of these factors into a consistent total valuation before policies on increased recycling of paper products are implemented on a large scale.

Energy use emerges as the most important sector in the environmental impacts of recycling. The possible use of waste paper for energy production adds a credit which should be compared with

Table 9.1. Results of calculations for conditions prevailing in the late 1980s and for increased recycling rates.

Energy consumption		*Emissions – Water*	
Electric power	Decreased	TSS	Increased
Heat and steam	Decreased	BOD	Increased
Fossil fuels	Increased	COD	Decreased
Nonrenewable primary		AOX	Decreased
energy sources	Increased		
Renewable primary		*Materials*	
energy sources	Decreased	Raw materials for	
		pulp and paper	
		production	
Emissions – Air		(other than wood)	Decreased
SO_2	Increased	Wood consumption	Decreased
NO_x	Increased		
CH_4	Decreased	*Waste production*	
Gross CO_2	Decreased	Gross solid waste	Decreased
CO	Decreased	Net solid waste	Increased
Net CO_2	Increased		
	(or decreased	*Forest management*	
	fixation)	Intensity	Decreased

large-scale fiber recycling and with the energy input necessary for collection, sorting, and re-pulping. The interactions between different economic sectors may also introduce problems if the potential heat value of household waste decreases. By not distinguishing between nonrenewable and renewable energy sources in the policy discussions on paper recycling, there is a risk that policy makers and consumers will get a distorted view of the impacts. In fact, as the calculations show, recycling would add to the consumption of nonrenewable energy sources under present conditions.

An important policy question which requires more consideration is the production of energy by burning waste paper. In the proposed regulations for Western Europe on paper recycling the burning alternative has been considered, but has often been ruled out. The results of this study indicate that burning waste paper for energy production may be a good alternative from the environmental and the economic points of view. This is especially true if some of the chemicals and heavy metals used in the pulp, paper, and printing processes were replaced with other materials. In this case, waste paper could be classified as a clean fuel.

Special concern in the policy process on a suitable rate of re-cycling must be directed toward the forest resources and forest management. Based on the calculations for 1989 in Western Europe it can be concluded that a dramatically increased recycling rate (the maximum recycling scenario) will probably generate the following effects:

- A decrease in total wood consumption of about 25% in the region.
- A decrease of the stumpage prices (in Western Europe and, indirectly, in other regions), resulting in less intensive forest management.
- A particularly decreased thinning intensity: it is estimated that about 50% of the thinning requirements may not be carried out.
- A revision of the forest policies in several countries toward less intensive policies than those currently in effect.
- Decreased stumpage prices creating the risk of a decreased afforestation rate in Western Europe and, indirectly, in other countries.
- A strong decrease in disposal possibilities for sawmill chips. The market for sawmill chips seems to decrease by more than 40% in the case of maximum recycling in Western Europe, in comparison with the figures for production in 1989. The in-come from chips is, in most cases, crucial to the survival of the sawmills.

These conclusions are based on the conditions in 1989. If we look into the future, the picture becomes more complicated. Nilsson *et al.* (1992) estimate that, if no radical measures are taken against current emissions of air pollutants in Europe, the long-term sus-tainable wood supply in the region will be approximately only 355 million cubic meters s.o.b. (including transfer of primary residues), instead of the 402.5 million cubic meters s.o.b. discussed earlier. The consumption of paper and board in the region is also estimated to increase by between 2% and 2.5% per year during the next 10 years (Jaakko Pöyry, 1992). These developments will tighten the wood balance, and as a result our conclusions on the effects of recycling on forest management may change. The long-term development and effects of recycling of paper products is a central component of the planned full-scale study for Western Europe.

The zero recycling scenario seems to generate the largest net fixation of carbon of the scenarios studied. However, the carbon balances produced have not been able to illustrate the dynamics of carbon fixation over time by forest growth. It is intended that these dynamics will be analyzed in the planned full-scale recycling study. As illustrated in the wood balances, the wood supply/demand balance will, at a strongly increased recycling rate, require that policy makers implement policies which stimulate wood consumption for energy production in Western Europe. Otherwise, there will be limited possibilities of carrying out satisfactory silviculture measures (thinnings) in the future in order to create vital forests.

When planning the collection and recycling of waste paper it is necessary to consider the different qualities of paper. High-grade waste paper can be recirculated and used for many different lower qualities, but low-grade waste paper can be used for only a few less-demanding purposes. Paper cannot be recirculated without additional primary fibers because a certain quality degradation will occur at each reuse cycle. Source separation of waste paper into different qualities should make the usage of recycled fibers in the paper production process more economical and should improve the allocation of waste paper qualities most suitable for energy production. However, the availability of different waste paper grades for different recycling schemes is not sufficiently known or documented for a generally acceptable evaluation of the environmental impacts of paper recycling. There are also large unquantified variations in energy demand and waste sludge generated by reused pulp production from different grades of waste paper.

We have not been able to analyze several, important policy questions in this study. Wiseman (1990) has pointed out that an increased recycling rate will also decrease the prices of final industrial forest products in the long-term due to decreased stumpage prices. This may lead to relatively higher consumption of paper products in the future due to their substitution for other products (like plastics) at decreased prices. A different scenario is also possible. Probability is high that the regulations proposed in Western Europe may cause large practical and economic problems when they are implemented. This may lead to the substitution of paper and board products for other products. Furthermore, there may be a risk that this replacement will take the form of more environmentally harmful products.

We have pointed out earlier that the current sites for the production of pulp for paper and the paper consumption sites do not coincide. An increased recycling rate will probably result in the relocation of the pulp and paper industry (the market for waste paper) to the sites where paper is consumed. Efforts to relocate the industry will incur considerable investment costs, and will have specific environmental impacts which we have not been able to analyze in this feasibility study.

An increased recycling rate may change the trade patterns of paper and board products in Western Europe (in addition to the changes caused by the relocation of industry). The net importing countries may be forced to import a higher amount of paper and board products with high degrees of primary fiber in order to obtain the necessary high-quality fibers for recycling. A changed trade pattern may also have environmental impacts.

When considering the uses of paper products it should be noted that they represent one of the most-used, renewable material, and energy sources of the world. Well-planned schemes for the recycling and energy recovery of used fibers and efficient forest-management programs are important factors in attaining sustainable development of the environment and the forest resources in Western Europe.

Appendix A: Production Data for the Maximum Recycling Scenario (M)

	Unit	Rate, unit	Rate, mass
PAPER & BOARD CONSUMPTION/CE			
Input			
Paper & board/ce	kg	716290.00007	716290.00007
Paper & board/imported/ce	kg	418360.00004	418360.00004
Output			
Paper & board consumption/ce	pcs	−1.00000	0.00000
Waste paper/collected/ce	kg	−645500.00006	−645500.00006
Waste paper/sewage	kg	−28820.00000	−28820.00000
Waste paper & board	kg	−136310.00001	−136310.00001
Paper & board/long term	kg	−98180.00001	−98180.00001
Paper & board/exported/ce	kg	−225860.00002	−225860.00002
Net input–output balance			20.00000
PAPER & BOARD CONSUMPTION/SC			
Input			
Paper & board/sc	kg	384090.00004	384090.00004
Paper & board/imported/sc	kg	5950.00000	5950.00000
Output			
Paper & board consumption/sc	pcs	−1.00000	0.00000
Waste paper/collected/sc	kg	−61170.00001	−61170.00001
Waste paper/sewage	kg	−3850.00000	−3850.00000
Waste paper & board	kg	−15310.00000	−15310.00000
Paper & board/long term	kg	−10880.00000	−10880.00000
Paper & board/exported/sc	kg	−298840.00003	−298840.00003
Net input–output balance			10.00000
PAPER & BOARD PRODUCTION/CE			
Input			
Bleached pulp/sulfate	kg	104359.99978	104359.99978
Brown pulp/sulfate	kg	69340.00005	69340.00005
Bleached pulp/sulfite	kg	26459.99972	26459.99972
Bleached pulp/tmp	kg	45719.99992	45719.99992
Brown pulp/reuse/ce	kg	401119.99975	401119.99975
Other material/paper & board	kg	69290.00014	69290.00014
Electric power	kWh	479910.00016	0.00000
Steam	MJ	5016050.42036	1790730.00007
Output			
Paper & board/ce	kg	−716290.00007	−716290.00007
Process water	kg	−1790730.00060	−1790730.00060
Net input–output balance			0.00125
PAPER & BOARD PRODUCTION/SC			
Input			
Bleached pulp/sulfate	kg	47110.00010	47110.00010

CE denotes Central Europe. SC denotes Scandinavia.

	Unit	Rate, unit	Rate, mass
Brown pulp/sulfate	kg	27989.9999	27989.99990
Bleached pulp/sulfite	kg	6180.00003	6180.00003
Bleached pulp/tmp	kg	56859.99993	56859.99993
Brown pulp/reuse/sc	kg	215090.00018	215090.00018
Other material/paper & board	kg	30850.00011	30850.00011
Electric power	kWh	257340.00005	0.00000
Steam	MJ	2689691.87699	960220.00009
Output			
Paper & board/sc	kg	−384090.00004	−384090.00004
Process water	kg	−960220.00001	−960220.00001
Net input–output balance			9.99971

PULP PROD/BLEACHED/BIRCH
Input

	Unit	Rate, unit	Rate, mass
Natural gas	kg	2526.67107	2526.67107
Process air	kg	2600563.73833	2600563.73833
Logs	kg	495502.14444	495502.14444
CaO/process	kg	943.65810	943.65810
NaOH/process	kg	2746.15110	2746.15110
Process water	kg	5942509.81162	5942509.81162
H_2SO_4/process	kg	593.76240	593.76240
Na_2SO_4/process	kg	593.76240	593.76240
$NaClO_3$/process	kg	3244.48740	3244.48740
HCl/process	kg	805.82040	805.82040
Methanol	kg	212.05800	212.05800
SO_2/process	kg	349.89570	349.89570
$MgSO_4$/process	kg	0.00000	0.00000
O_2/process	kg	2099.37420	2099.37420
Electric power	kWh	58528.00795	0.00000
Steam	MJ	925633.16937	330451.04147
Output			
Wastewater	kg	−6079914.91382	−6079914.91382
NO_x	kg	−100.02034	0.00000
TSP	kg	−38.17044	0.00000
Pine oil/raw	kg	0.00000	0.00000
Terpentine/raw	kg	0.00000	0.00000
BOD	kg	−200.39481	0.00000
COD	kg	−4135.13100	0.00000
AOX	kg	−53.01450	0.00000
Phosphorus	kg	−5.30145	0.00000
Na_2SO_4/process	kg	−919.27143	−919.27143
NaCl/process	kg	−541.80819	−541.80819
Solid waste/other industrial	kg	−7422.02999	−7422.02999
Solid waste/other municipal	kg	−2120.58000	−2120.58000
Cl_2	kg	−7.33933	0.00000
TRS	kg	−84.82320	0.00000
SO_2	kg	−190.85220	0.00000

	Unit	Rate, unit	Rate, mass
CO_2	kg	–256791.63490	0.00000
Waste air	kg	–2744828.91620	–2744828.91620
Bleached pulp/sulfate	kg	–106028.99992	–106028.99992
TSS	kg	–233.26380	0.00000
Steam	MJ	–1236324.64627	–441367.89872
Electric power	kWh	–110660.11104	0.00000
Net input–output balance			2.04166

PULP PROD/BLEACHED/PINE
Input

Natural gas	kg	1580.89239	1580.89239
Process air	kg	1378259.60966	1378259.60966
Logs	kg	237111.13786	237111.13786
CaO/process	kg	489.85398	489.85398
NaOH/process	kg	1726.75800	1726.75800
Process water	kg	2555512.77362	2555512.77362
H_2SO_4/process	kg	299.91060	299.91060
Na_2SO_4/process	kg	354.43980	354.43980
$NaClO_3$/process	kg	1617.69960	1617.69960
HCl/process	kg	404.42490	404.42490
Methanol	kg	122.69070	122.69070
SO_2/process	kg	149.95530	149.95530
$MgSO_4$/process	kg	199.94040	199.94040
O_2/process	kg	1254.17160	1254.17160
Electric power	kWh	28082.53798	0.00000
Steam	MJ	481674.59966	171957.83208

Output

Bleached pulp/sulfate	kg	–45440.99996	–45440.99996
Waste air	kg	–1453158.64667	–1453158.64667
CO_2	kg	–136076.70967	0.00000
TSP	kg	–22.35697	0.00000
NO_x	kg	–54.25655	0.00000
SO_2	kg	–109.05840	0.00000
TRS	kg	–45.44100	0.00000
Cl_2	kg	–3.14452	0.00000
Solid waste/other municipal	kg	–908.82000	–908.82000
Solid waste/other industrial	kg	–3180.87000	–3180.87000
Wastewater	kg	–2618037.77220	–2618037.77220
TSS	kg	–99.97020	0.00000
BOD	kg	–118.14660	0.00000
COD	kg	–2385.65250	0.00000
AOX	kg	–72.70560	0.00000
Phosphorus	kg	–2.24933	0.00000
Pine oil/raw	kg	–1276.89210	–1276.89210
Terpentine/raw	kg	–90.88200	–90.88200
Na_2SO_4/process	kg	–463.49820	–463.49820
NaCl/process	kg	–268.10190	–268.10190

	Unit	Rate, unit	Rate, mass
Steam	MJ	−639263.07867	−228216.91909
Electric power	kWh	−57218.80226	0.00000
Net input–output balance			1.31164

PULP PROD/BROWN/SR
Input

	Unit	Rate, unit	Rate, mass
Natural gas	kg	2319.37390	2319.37390
Process air	kg	1820168.32921	1820168.32921
Heavy fuel oil	kg	324.10890	324.10890
Logs	kg	394673.14979	394673.14979
CaO/process	kg	767.93370	767.93370
NaOH/process	kg	475.94370	475.94370
Process water	kg	1145546.84698	1145546.84698
H_2SO_4/process	kg	541.15480	541.15480
Na_2SO_4/process	kg	432.14520	432.14520
Electric power	kWh	45647.76998	0.00000
Steam	MJ	837037.99964	298822.56587

Output

	Unit	Rate, unit	Rate, mass
Brown pulp/sulfate	kg	−97329.99995	−97329.99995
Waste air	kg	−1919947.15176	−1919947.15176
CO_2	kg	−179768.50990	0.00000
TSP	kg	−26.82025	0.00000
NO_x	kg	−71.44022	0.00000
SO_2	kg	−136.26200	0.00000
TRS	kg	−58.39800	0.00000
Wastewater	kg	−1268209.89941	−1268209.89941
TSS	kg	−38.93200	0.00000
BOD	kg	−48.66500	0.00000
COD	kg	−1021.96500	0.00000
Phosphorus	kg	−9.63567	0.00000
Pine oil/raw	kg	−2703.82740	−2703.82740
Terpentine/raw	kg	−194.66000	−194.66000
Solid waste/other municipal	kg	−1946.60000	−1946.60000
Steam	MJ	−1046888.29253	−373739.12043
Electric power	kWh	−93704.18707	0.00000
Net input–output balance			0.29310

PULP PROD/BLEACHED/SULFITE
Input

	Unit	Rate, unit	Rate, mass
Logs	kg	154436.15885	154436.15885
MgO/process	kg	447.16800	447.16800
NaOH/process	kg	1302.33599	1302.33599
O_2/process	kg	512.44800	512.44800
Auxiliary chemicals	kg	793.15199	793.15199
SO_2/process	kg	1175.03999	1175.03999
Process water	kg	1474567.15064	1474567.15064
Electric power	kWh	18858.66653	0.00000
Steam	MJ	378623.99722	135168.76701

	Unit	Rate, unit	Rate, mass
H_2O_2/process	kg	362.30400	362.30400
Process air	kg	382377.59719	382377.59719
Output			
Acetic acid	kg	−3035.51998	−3035.51998
Furfural	kg	−815.99999	−815.99999
Waste air	kg	−484527.74040	−484527.74040
CO_2	kg	−103305.59923	0.00000
TSP	kg	−42.43200	0.00000
NO_x	kg	−92.69760	0.00000
SO_2	kg	−112.93440	0.00000
Solid waste/other industrial	kg	−2970.23998	−2970.23998
Wastewater	kg	−1403519.98970	−1403519.98970
COD	kg	−228.48000	0.00000
BOD5	kg	−32.64000	0.00000
AOX	kg	−3.26400	0.00000
TSS	kg	−42.43200	0.00000
Bleached pulp/sulfite	kg	−32639.99976	−32639.99976
Sulfate	kg	−1207.67999	0.00000
TOC	kg	−81.60000	0.00000
Steam	MJ	−622444.79544	−222212.79197
Electric power	kWh	−23482.66648	0.00000
Na_2CO_3/process	kg	−1419.83999	−1419.83999
Net input–output balance			0.00011
PULP PROD/BLEACHED/TMP			
Input			
Electric power	kWh	284944.67197	0.00000
Steam	MJ	399197.10916	142513.36797
Logs	kg	221349.17527	221349.17527
Process water	kg	2849444.66812	2849444.66812
H_2O_2/process	kg	2393.19140	2393.19140
NaOH/process	kg	1937.73620	1937.73620
Silicate	kg	2393.19140	2393.19140
DTPA	kg	455.45520	455.45520
Output			
Bleached pulp/tmp	kg	−102579.99985	−102579.99985
Wastewater	kg	−3107226.15613	−3107226.15613
COD	kg	−6496.39139	0.00000
BOD	kg	−5128.99999	0.00000
TSS	kg	−570.34480	0.00000
Solid waste/other municipal	kg	−10680.62958	−10680.62958
Net input–output balance			0.00002
PULP PROD/BROWN/REUSE/CE			
Input			
Waste paper/supply/ce	kg	451203.84292	451203.84292
Process water	kg	5348265.32630	5348265.32630

	Unit	Rate, unit	Rate, mass
Electric power	kWh	189416.88628	0.00000
Steam	MJ	449434.17328	160447.99986
Output			
Brown pulp/reuse/ce	kg	–401119.99975	–401119.99975
Wastewater	kg	–5528324.08299	–5528324.08299
BOD	kg	–6686.67040	0.00000
COD	kg	–5792.17280	0.00000
TSS	kg	–2230.22720	0.00000
Solid waste/other municipal	kg	–30473.08638	–30473.08638
Net input–output balance			0.00004

AUX MAT SUPPLY/PAPER & BOARD

Input			
Starch	kg	80112.00020	80112.00020
$CaCO_3$/process	kg	10014.00002	10014.00002
TiO_2	kg	10014.00002	10014.00002
Output			
Other material/paper & board	kg	–100140.00024	–100140.00024
Net input–output balance			0.00000

PULP PROD/BROWN/REUSE/SC

Input			
Waste paper/supply/sc	kg	241946.13761	241946.13761
Process water	kg	2867865.95215	2867865.95215
Electric power	kWh	101569.79989	0.00000
Steam	MJ	240997.19903	86036.00005
Output			
Brown pulp/reuse/sc	kg	–215090.00018	–215090.00018
Wastewater	kg	–2964417.70233	–2964417.70233
BOD	kg	–3585.55030	0.00000
COD	kg	–3105.89960	0.00000
TSS	kg	–1195.90040	0.00000
Solid waste/other municipal	kg	–16340.38731	–16340.38731
Net input–output balance			0.00002

GAS PIPELINE DISTRIBUTION

Input			
Natural gas/cleaned	m^3	10038.45417	7428.45608
Natural gas	kg	14.24000	14.24000
Process air	kg	4114.94019	4114.94019
Output			
Natural gas	kg	–7074.72008	–7074.72008
VOC	kg	–33.95866	0.00000
NO_x	kg	–391.93949	0.00000
CH_4	kg	–151.39901	0.00000
CO	kg	–49.52304	0.00000
Waste air	kg	–4482.96713	–4482.96713
Net input–output balance			0.05094

	Unit	Rate, unit	Rate, mass
HARVESTING			
Input			
Wood biomass	kg	2312196.63104	2312196.63104
Light fuel oil	kg	284.48489	284.48489
Middle distillate	kg	3807.15903	3807.15903
Process air	kg	72203.17567	72203.17567
Output			
Logs	kg	−1503071.76620	−1503071.76620
Harvesting wastes	kg	−809125.76669	−809125.76669
Waste air	kg	−76290.43213	−76290.43213
CO	kg	−314.26826	0.00000
VOC	kg	−187.92305	0.00000
NO_x	kg	−210.16250	0.00000
TSP	kg	−16.36394	0.00000
Net input–output balance			3.48562
LIME (CaO) MANUFACTURE			
Input			
$CaCO_3$/process	kg	3705.03325	3705.03325
Electric power	kWh	784.59528	0.00000
Heat	MJ	9690.76432	0.00000
Process air	kg	7110.66986	7110.66986
Output			
CaO/process	kg	−2201.44578	−2201.44578
CO_2	kg	−1430.93976	0.00000
Waste air	kg	−7110.66986	−7110.66986
Net input–output balance			1503.58747
NaOH/DIAPHRAGM PROCESS			
Input			
NaCl/process	kg	13433.91532	13433.91532
H_2SO_4/process	kg	436.43640	436.43640
HCl/process	kg	414.62702	414.62702
Cooling water	l	1229949.58003	1229949.58003
Steam	MJ	2241.30873	800.14722
Process water	kg	65593.99392	65593.99392
$CaCl_2$/process	kg	94.86666	94.86666
Electric power	kWh	22.14108	0.00000
Na_2CO_3/process	kg	580.47782	580.47782
Heat	MJ	72642.65319	0.00000
Output			
NaOH/process	kg	−8292.54032	−8292.54032
H_2SO_4/process	kg	−580.47782	−580.47782
H_2/process	kg	−218.25966	−218.25966
Cooling water	l	−1229949.58003	−1229949.58003
Wastewater	kg	−62774.53021	−62774.53021
Solid waste/other industrial	kg	−1163.85803	−1163.85803

	Unit	Rate, unit	Rate, mass
Cl_2	kg	−0.58189	−0.58189
H_2	kg	−27.28246	−27.28246
TDS	kg	−400.11507	0.00000
NaOH	kg	−7.96084	0.00000
Chlorides	g	−7297.43548	0.00000
Sulfates	kg	−4.00074	0.00000
Asbestos	g	−72.72558	0.00000
Cl_2/process	kg	−7297.43548	−7297.43548
Net input–output balance			999.49847
H_2SO_4 PRODUCTION			
Input			
S_2/process	kg	423.03287	423.03287
Catalyst	kg	0.00000	0.00000
Cooling water	l	95527.71659	95527.71659
Process water	kg	735.82160	735.82160
Output			
H_2SO_4/process	kg	−1290.91509	−1290.91509
Steam	MJ	−2284.91971	−815.71634
Cooling water	l	−95527.71659	−95527.71659
SO_2	kg	−10.65005	−10.65005
H_2SO_4	kg	−0.45182	−0.45182
Sulfates	kg	−0.22591	−0.22591
Net input–output balance			959.10473
SODIUM CHLORATE PRODUCTION			
Input			
Electric power	kWh	28395.17206	0.00000
Heat	MJ	13069.55865	0.00000
NaCl/process	kg	2722.82472	2722.82472
Barium chloride	kg	36.46640	36.46640
HCl/process	kg	121.55467	121.55467
NaOH/process	kg	97.24374	97.24374
Sodium dichromate	kg	1.45866	1.45866
Process water	kg	8265.71789	8265.71789
Na_2CO_3/process	kg	17.99009	17.99009
Output			
$NaClO_3$/process	kg	−4862.18700	−4862.18700
Wastewater	kg	−6401.06918	−6401.06918
Net input–output balance			0.00000
METHANOL PRODUCTION			
Input			
Electric power	kWh	11.04671	0.00000
Cooling water	l	19817.12302	19817.12302
Natural gas	kg	251.06152	251.06152
CO_2/process	kg	130.55199	130.55199
Process water	kg	512.16551	512.16551
Heat/natural gas	MJ	630.66655	0.00000

	Unit	Rate, unit	Rate, mass
Output			
Methanol	kg	−334.74870	−334.74870
Acids	g	−6.69497	0.00000
VOC	kg	−46.53007	−46.53007
TSP	kg	−0.33475	−0.33475
Dimethylether	kg	−30.12738	−30.12738
Cooling water	l	−19817.12302	−19817.12302
Wastewater	kg	−512.16551	−512.16551
By-products	kg	−46.53007	−46.53007
Net input–output balance			76.65745
O$_2$ PRODUCTION			
Input			
Process air	kg	16701.09319	16701.09319
Cooling water	l	121005.60574	121005.60574
Electric power	kWh	1670.92118	0.00000
Output			
O$_2$/process	kg	−3865.99379	−3865.99379
N$_2$	kg	−1198.45808	−1198.45808
Ar	kg	−77.31988	−77.31988
N$_2$/impure	kg	−11559.32144	−11559.32144
Wastewater	kg	−121005.60574	−121005.60574
Net input–output balance			0.00000
FUEL DISTRIB/HEAVY FUEL OIL			
Input			
Fuel	kg	339.77075	339.77075
Output			
Heavy fuel oil	kg	−337.07416	−337.07416
Oils & greases	kg	−2.02244	−2.02244
VOC	kg	−0.67415	−0.67415
Net input–output balance			0.00000
H$_2$O$_2$ PRODUCTION			
Input			
H$_2$/process	kg	166.23345	166.23345
Ethyl anthraquinone	kg	9.67981	9.67981
Trioctyl phosphate	kg	7.46728	7.46728
Cylcohexyl pyrrolidone	kg	7.46728	7.46728
Cyclosol 63	kg	19.05539	19.05539
Xylene	kg	11.03498	11.03498
Sodium pyrophosphate	kg	2.76566	2.76566
Catalyst/Pd	kg	1.05095	1.05095
Process water	kg	16040.82252	16040.82252
Steam	MJ	46481.66477	16593.95432
Electric power	kWh	4756.93357	0.00000
Output			
H$_2$O$_2$/process	kg	−2765.65905	−2765.65905

	Unit	Rate, unit	Rate, mass
Wastewater	kg	−30093.85524	−30093.85524
TDS	kg	−58.52135	0.00000
Net input–output balance			0.01734
WASTE PAPER SUPPLY/REUSE/CE			
Input			
Waste paper/collected/ce	kg	451203.84292	451203.84292
Waste paper/imported/sc/ce	kg	0.00000	0.00000
Waste paper/imported/ow/ce	kg	0.00000	0.00000
Output			
Waste paper/supply/ce	kg	−451203.84292	−451203.84292
Net input–output balance			0.00000
STARCH PRODUCTION			
Input			
Potato	kg	320448.00078	320448.00078
Steam	MJ	63128.25615	22536.78745
Output			
Starch	kg	−80112.00020	−80112.00020
Potato fiber	kg	−12817.92003	−12817.92003
Wastewater	kg	−227518.08055	−227518.08055
TDS	kg	−3204.48001	0.00000
Net input–output balance			22536.78745
LIMESTONE MINING & PREPARATION			
Input			
Limestone rock	kg	15554.63992	15554.63992
Process air	kg	14337.76167	14337.76167
Process water	kg	14404.98493	14404.98493
Electric power	kWh	787.19813	0.00000
Heat	MJ	1566.71360	0.00000
Explosive	kg	13.71903	13.71903
Output			
$CaCO_3$/process	kg	−13719.03327	−13719.03327
Wastewater	kg	−15701.43358	−15701.43358
Waste air	kg	−14878.29158	−14878.29158
TSP	kg	−385.23045	0.00000
Net input–output balance			12.34713
TITANIUM DIOXIDE PRODUCTION			
Input			
Electric power	kWh	275.28486	0.00000
Heat	MJ	107720.59826	0.00000
O_2/liquid	kg	4506.30001	4506.30001
Cl_2/process	kg	1502.10000	1502.10000
Rutile	kg	11315.82003	11315.82003
Output			
TiO_2	kg	−10014.00002	−10014.00002
Net input–output balance			7310.22002

	Unit	Rate, unit	Rate, mass
WASTE PAPER SUPPLY/REUSE/SC			
Input			
Waste paper/collected/sc	kg	57760.00176	57760.00176
Waste paper/imported/ce/sc	kg	183270.20979	183270.20979
Waste paper/imported/ow/sc	kg	915.92606	915.92606
Output			
Waste paper/supply/sc	kg	–241946.13761	–241946.13761
Net input–output balance			0.00000
NATURAL GAS WELL & PROCESSING			
Input			
Natural gas/crude	m^3	10419.91543	7710.73741
Electric power	kWh	637.24107	0.00000
Middle distillate	kg	151.23345	151.23345
Light fuel oil	kg	32.17777	32.17777
Natural gas	kg	297.58395	297.58395
Process air	kg	13943.81438	13943.81438
Output			
Natural gas/cleaned	m^3	–10038.45417	–7428.45608
TDS	kg	–2.07415	–2.07415
VOC	kg	–3.41307	0.00000
CH$_4$	kg	–27.70613	0.00000
SO$_2$	kg	–22.48614	0.00000
Waste air	kg	–14706.33535	–14706.33535
Oils & greases	kg	–0.49510	–0.49510
Net input–output balance			1.81372
WOOD BIOMASS FORMATION			
Input			
CO$_2$	kg	1834427.44117	1834427.44117
H$_2$O	kg	1811906.64599	1811906.64599
Output			
Wood biomass	kg	–2312196.63104	–2312196.63104
O$_2$	kg	–1334137.45611	–1334137.45611
Net input–output balance			0.00000
FUEL DISTRIB/LIGHT FUEL OIL			
Input			
Fuel	kg	319.09038	319.09038
Output			
Light fuel oil	kg	–317.81910	–317.81910
VOC	kg	–1.27128	–1.27128
Net input–output balance			0.00000
FUEL DISTRIB/MIDDLE DISTILLATE			
Input			
Fuel	kg	6403.75750	6403.75750
Output			
Middle distillate	kg	–6352.93403	–6352.93403

	Unit	Rate, unit	Rate, mass
Oils & greases	kg	−12.70587	−12.70587
VOC	kg	−38.11760	−38.11760
Net input–output balance			0.00000

SALT MINING & PREPARATION
Input

	Unit	Rate, unit	Rate, mass
Heat	MJ	3621.85187	0.00000
Electric power	kWh	1442.60201	0.00000
Salt rock	kg	17361.86872	17361.86872

Output

	Unit	Rate, unit	Rate, mass
NaCl/process	kg	−15346.82994	−15346.82994
Solid waste/other industrial	kg	−2008.90004	−2008.90004
TSP	kg	−4.60405	−4.60405
Acids	g	−1227.74640	−1.22775
Heavy metals	kg	−0.30694	−0.30694
Net input–output balance			0.00000

Na_2CO_3/SOLVAY PROCESS
Input

	Unit	Rate, unit	Rate, mass
NaCl/process	kg	0.00000	0.00000
$CaCO_3$/process	kg	0.00000	0.00000
NH_3/process	kg	0.00000	0.00000
Derived coal	kg	0.00000	0.00000
Process air	kg	0.00000	0.00000
Process water	kg	0.00000	0.00000
Electric power	kWh	0.00000	0.00000
Heat	MJ	0.00000	0.00000
CO_2/process	kg	0.00000	0.00000

Output

	Unit	Rate, unit	Rate, mass
Na_2CO_3/process	kg	0.00000	0.00000
CO_2/process	kg	0.00000	0.00000
NH_3/process	kg	0.00000	0.00000
Wastewater	kg	0.00000	0.00000
Waste air	kg	0.00000	0.00000
NH_3	kg	0.00000	0.00000
TSP	kg	0.00000	0.00000
Net input–output balance			0.00000

OIL REFINING/FUEL
Input

	Unit	Rate, unit	Rate, mass
Crude oil	kg	7557.00194	7557.00194
Salt water	kg	26837.95080	26837.95080
Process water	kg	2825.04745	2825.04745
NaOH/process	kg	4.94383	4.94383
Catalyst	kg	0.07063	0.07063
Process air	kg	6804.83305	6804.83305

Output

	Unit	Rate, unit	Rate, mass
Waste air	kg	−7288.55180	−7288.55180

	Unit	Rate, unit	Rate, mass
Wastewater	kg	−29668.01271	−29668.01271
Oils & greases	kg	−0.07063	0.00000
Phenols	kg	−0.00706	0.00000
Sulfides	kg	−0.01059	0.00000
CO_2	kg	−1412.52373	0.00000
SO_2	kg	−17.65655	0.00000
NO_x	kg	−2.47192	0.00000
VOC	kg	−5.29696	0.00000
CH_4	kg	−0.70626	0.00000
Solid waste/other industrial	kg	−10.59393	−10.59393
Fuel	kg	−7062.61863	−7062.61863
Net input–output balance			0.07063
H_2 PRODUCTION			
Input			
Natural gas	kg	0.00000	0.00000
Process air	kg	0.00000	0.00000
Steam	MJ	0.00000	0.00000
Heat	MJ	0.00000	0.00000
Cooling water	l	0.00000	0.00000
Output			
H_2/process	kg	0.00000	0.00000
S_2/process	kg	0.00000	0.00000
Waste air	kg	0.00000	0.00000
Cooling water	l	0.00000	0.00000
CO_2/process	kg	0.00000	0.00000
Net input–output balance			0.00000
ETHYL ANTHRAQUINONE PRODUCTION			
Input			
Electric power	kWh	4.16226	0.00000
Steam	MJ	352.48121	125.83579
H_2O_2/process	kg	10.16366	10.16366
Phenol	kg	22.26326	22.26326
Phosphoric acid	kg	0.05517	0.05517
Isopropyl ether	kg	0.25167	0.25167
NaOH/process	kg	0.24199	0.24199
Sodium bisulfite	kg	0.00774	0.00774
Perchloric acid	kg	0.03775	0.03775
Process water	kg	54.20619	54.20619
Output			
Ethyl anthraquinone	kg	−9.67968	−9.67968
Catechol	kg	−9.67968	−9.67968
Phenol	kg	−12.58358	−12.58358
Isopropyl ether	kg	−0.22650	−0.22650
Wastewater	kg	−180.89379	−180.89379
TDS	kg	−0.85181	0.00000
Net input–output balance			0.00000

	Unit	Rate, unit	Rate, mass
XYLENE PRODUCTION			
Input			
Aromatics output	kg	33.51658	33.51658
Output			
Xylene	kg	−11.03483	−11.03483
By-products	kg	−22.48175	−22.48175
Net input–output balance			0.00000
WASTE PAPER EXPORT/SC			
Input			
Waste paper/surplus/sc	kg	0.00000	0.00000
Output			
Waste paper/imported/sc/ce	kg	0.00000	0.00000
Net input–output balance			0.00000
POTATO CULTIVATION			
Input			
Potato	kg	31692.65928	31692.65928
Nitrogen fertilizer	kg	1320.52747	1320.52747
Phosphorus fertilizer	kg	264.10549	264.10549
Potassium fertilizer	kg	1760.70329	1760.70329
Herbicides	kg	40.49618	40.49618
Fungicides	kg	30.81231	30.81231
Tractor hour	h	440.17582	0.00000
CO_2	kg	113913.27740	113913.27740
H_2O	kg	289313.83520	289313.83520
Output			
Potato	kg	−352140.65868	−352140.65868
Leachate	kg	−2589.11419	−2589.11419
O_2	kg	−82846.01992	−82846.01992
CH_4	kg	−202.83302	−202.83302
CO_2	kg	−557.79080	−557.79080
Net input–output balance			0.00000
O_2 LIQUEFACTION			
Input			
O_2	kg	4515.54800	4515.54800
Electric power	kWh	2527.63541	0.00000
Output			
O_2/liquid	kg	−4506.30001	−4506.30001
Net input–output balance			9.24799
WASTE PAPER EXPORT/CE			
Input			
Waste paper/surplus/ce	kg	183270.20979	183270.20979
Output			
Waste paper/imported/ce/sc	kg	−183270.20979	−183270.20979
Net input–output balance			0.00000

	Unit	Rate, unit	Rate, mass
NH₃ PRODUCTION			
Input			
Natural gas	kg	0.00000	0.00000
Process air	kg	0.00000	0.00000
Process water	kg	0.00000	0.00000
Cooling water	l	0.00000	0.00000
Electric power	kWh	0.00000	0.00000
Heat	MJ	0.00000	0.00000
Output			
NH_3/process	kg	0.00000	0.00000
CO_2/process	kg	0.00000	0.00000
Waste air	kg	0.00000	0.00000
Wastewater	kg	0.00000	0.00000
Steam	MJ	0.00000	0.00000
NO_x	kg	0.00000	0.00000
NH_3/air	kg	0.00000	0.00000
CO	kg	0.00000	0.00000
H_2	kg	0.00000	0.00000
NH_3	kg	0.00000	0.00000
Mono-ethyl amine	kg	0.00000	0.00000
Cooling water	l	0.00000	0.00000
Net input–output balance			0.00000
DERIVED COAL PRODUCTION			
Input			
Electric power	kWh	0.00000	0.00000
Heat	MJ	0.00000	0.00000
Hard coal	kg	0.00000	0.00000
Process air	kg	0.00000	0.00000
Process water	kg	0.00000	0.00000
Output			
Derived coal	kg	0.00000	0.00000
SO_2	kg	0.00000	0.00000
NO_x	kg	0.00000	0.00000
CO	kg	0.00000	0.00000
BOD5	kg	0.00000	0.00000
Cd	g	0.00000	0.00000
Cyanides	kg	0.00000	0.00000
Coal tar	kg	0.00000	0.00000
Light fuel oil/impure	kg	0.00000	0.00000
TSP	kg	0.00000	0.00000
VOC	kg	0.00000	0.00000
Wastewater	kg	0.00000	0.00000
Derived coal/breeze	kg	0.00000	0.00000
Gas/derived coal production	kg	0.00000	0.00000
Water vapor	kg	0.00000	0.00000
Process water/used	kg	0.00000	0.00000

	Unit	Rate, unit	Rate, mass
NH_3/process	kg	0.00000	0.00000
CO_2	kg	0.00000	0.00000
NH_3	kg	0.00000	0.00000
Oils & greases	kg	0.00000	0.00000
TDS	kg	0.00000	0.00000
Net input–output balance			0.00000
OIL DISTRIBUTION			
Input			
Oil/extracted	kg	7922.72645	7922.72645
Natural gas	kg	25.09286	25.09286
Process air	kg	489.62449	489.62449
Output			
Crude oil	kg	–7922.72645	–7922.72645
Waste air	kg	–501.82549	–501.82549
Net input–output balance			12.89186
PHENOL/ACETONE VIA CUMENE			
Input			
Benzene	kg	10.18496	10.18496
Propylene	kg	5.72069	5.72069
NaOH/process	kg	0.95442	0.95442
H_2SO_4/process	kg	0.13455	0.13455
Cooling water	l	2129.52882	2129.52882
Process water	kg	8.32452	8.32452
Steam	MJ	126.02939	44.99249
Electric power	kWh	3.43367	0.00000
Heat/middle distillate	MJ	59.23962	0.00000
Output			
Phenol	kg	–9.67968	–9.67968
Acetone	kg	–5.87266	–5.87266
Heavy ends	kg	–0.63595	–0.63595
Process water	kg	–8.32452	–8.32452
Steam/condensed	kg	–45.01050	–45.01050
Cooling water	l	–2129.52882	–2129.52882
Net input–output balance			0.78831
AROMATICS PRODUCTION			
Input			
Additives	kg	0.01120	0.01120
Cooling water	l	2315.40764	2315.40764
Steam	MJ	225.17975	80.38917
Electric power	kWh	0.56130	0.00000
Natural gas	kg	0.34414	0.34414
Gasoline/cracking	kg	62.82133	62.82133
Output			
Process water	kg	–80.40317	–80.40317
Aromatics output	kg	–55.88274	–55.88274
Cooling water	l	–2315.40764	–2315.40764

	Unit	Rate, unit	Rate, mass
By-products	kg	−6.93859	−6.93859
Net input–output balance			0.34134

NITROGEN FERTILIZER PRODUCTION
Input

Electric power	kWh	17166.75582	0.00000
Nitrogen compounds	kg	1320.51968	1320.51968

Output

Nitrogen fertilizer	kg	−1320.51968	−1320.51968
NH₃	kg	−1.71668	−1.71668
Net input–output balance			1.71668

PHOSPHORUS FERTILIZER PRODUCTION
Input

Steam	MJ	8451.32594	3017.12336
Phosphorus compounds	kg	264.10394	264.10394

Output

Phosphorus fertilizer	kg	−264.10394	−264.10394
Process water	kg	−3016.06694	−3016.06694
Net input–output balance			1.05642

POTASSIUM FERTILIZER PRODUCTION
Input

Steam	MJ	17607.03217	6285.71048
Potassium compounds	kg	1760.70322	1760.70322

Output

Potassium fertilizer	kg	−1760.70322	−1760.70322
Process water	kg	−6285.71048	−6285.71048
Net input–output balance			0.00000

1 TRACTOR HOUR
Input

Middle distillate	kg	2394.54235	2394.54235
Process air	kg	35918.13525	35918.13525

Output

Tractor hour	h	−440.17323	0.00000
CO₂	kg	−1404.15259	0.00000
Waste air	kg	−38312.67760	−38312.67760
Net input–output balance			0.00000

COAL CLEANING/HARD
Input

Coal/run-of-mine	kg	0.00000	0.00000
Electric power	kWh	0.00000	0.00000
Process water	kg	0.00000	0.00000
Process air	kg.	0.00000	0.00000

Output

Hard coal	kg	0.00000	0.00000
TSP	kg	0.00000	0.00000

	Unit	Rate, unit	Rate, mass
NO$_x$	kg	0.00000	0.00000
VOC	kg	0.00000	0.00000
TDS	kg	0.00000	0.00000
TSS	kg	0.00000	0.00000
Al	kg	0.00000	0.00000
NH$_4$	kg	0.00000	0.00000
Sulfates	kg	0.00000	0.00000
Solid waste/other industrial	kg	0.00000	0.00000
CO	kg	0.00000	0.00000
Waste air	kg	0.00000	0.00000
Wastewater	kg	0.00000	0.00000
Net input–output balance			0.00000
OIL WELL			
Input			
Heavy fuel oil	kg	12.96641	12.96641
Natural gas	kg	26.81139	26.81139
Crude oil/reservoir	t	8.19844	8198.43733
Electric power	kWh	1763.59891	0.00000
Salt water	kg	42782.72282	42782.72282
Light fuel oil	kg	1.16432	1.16432
Process air	kg	3324.45524	3324.45524
Output			
Oil/extracted	kg	−7922.72645	−7922.72645
Solid waste/other industrial	kg	−4.75364	−4.75364
VOC	kg	−0.79227	0.00000
TDS	kg	−100.22249	0.00000
CH$_4$	kg	−0.79227	0.00000
Wastewater	kg	−42884.21294	−42884.21294
Waste air	kg	−3534.88286	−3534.88286
Oils & greases	kg	−1.26764	0.00000
Net input–output balance			0.01837
BENZENE PRODUCTION			
Input			
Aromatics output	kg	22.36616	22.36616
Output			
Benzene	kg	−10.18496	−10.18496
By-products	kg	−12.18121	−12.18121
Net input–output balance			0.00000
PROPYLENE PRODUCTION			
Input			
Cracking output	kg	26.97391	26.97391
Output			
Propylene	kg	−5.72069	−5.72069
By-products	kg	−21.25323	−21.25323
Net input–output balance			0.00000

	Unit	Rate, unit	Rate, mass
GASOLINE/CRACKING PRODUCTION			
Input			
Cracking output	kg	238.41463	238.41463
Output			
Gasoline/cracking	kg	−62.82133	−62.82133
By-products	kg	−175.59329	−175.59329
Net input–output balance			0.00000
COAL MINING/SURFACE			
Input			
Coal seam	kg	0.00000	0.00000
Heat	MJ	0.00000	0.00000
Process air	kg	0.00000	0.00000
Process water	kg	0.00000	0.00000
Electric power	kWh	0.00000	0.00000
Middle distillate	kg	0.00000	0.00000
Output			
Coal/run-of-mine	kg	0.00000	0.00000
Waste air	kg	0.00000	0.00000
Wastewater	kg	0.00000	0.00000
TSP	kg	0.00000	0.00000
CH_4	kg	0.00000	0.00000
VOC	kg	0.00000	0.00000
SO_2	kg	0.00000	0.00000
NO_x	kg	0.00000	0.00000
TDS	kg	0.00000	0.00000
TSS	kg	0.00000	0.00000
Fe/diss	kg	0.00000	0.00000
Mn	g	0.00000	0.00000
Al	kg	0.00000	0.00000
NH_4	kg	0.00000	0.00000
Sulfates	kg	0.00000	0.00000
CO	kg	0.00000	0.00000
Solid waste/other industrial	kg	0.00000	0.00000
Net input–output balance			0.00000
COAL MINING/UNDERGROUND			
Input			
Electric power	kWh	0.00000	0.00000
Process water	kg	0.00000	0.00000
Coal seam	kg	0.00000	0.00000
Process air	kg	0.00000	0.00000
Output			
NH_4	kg	0.00000	0.00000
CH_4	kg	0.00000	0.00000
Chlorides	g	0.00000	0.00000
Coal/run-of-mine	kg	0.00000	0.00000
TDS	kg	0.00000	0.00000

	Unit	Rate, unit	Rate, mass
TSS	kg	0.00000	0.00000
Fe/susp	kg	0.00000	0.00000
Fe/diss	kg	0.00000	0.00000
Mn	g	0.00000	0.00000
Al	kg	0.00000	0.00000
Sulfates	kg	0.00000	0.00000
Wastewater	kg	0.00000	0.00000
Waste air	kg	0.00000	0.00000
Solid waste/other industrial	kg	0.00000	0.00000
TSP	kg	0.00000	0.00000
VOC	kg	0.00000	0.00000
SO_2	kg	0.00000	0.00000
NO_x	kg	0.00000	0.00000
CO	kg	0.00000	0.00000
Net input–output balance			0.00000

STEAMCRACKING
Input

	Unit	Rate, unit	Rate, mass
Naphtha	kg	341.11638	341.11638
Electric power	kWh	6.82233	0.00000
Cooling water	l	19443.63343	19443.63343
Process water	kg	170.55819	170.55819
Natural gas	kg	32.65848	32.65848
CH_4/process	kg	57.98978	57.98978
Process air	kg	555.18737	555.18737

Output

	Unit	Rate, unit	Rate, mass
TSP	kg	−0.00341	0.00000
CO	kg	−0.00887	0.00000
NO_x	kg	−0.13338	0.00000
SO_2	kg	−0.00341	0.00000
VOC	kg	−0.02729	0.00000
TDS	kg	−0.01262	0.00000
H_2/process	kg	−6.82233	−6.82233
Acetylene	kg	−1.70558	−1.70558
Residual fuel oil	kg	−8.52791	−8.52791
BOD5	kg	−0.05321	0.00000
COD7	kg	−0.16544	0.00000
Oils & greases	kg	−0.00887	0.00000
Cooling water	l	−19443.63343	−19443.63343
Wastewater	kg	−170.55819	−170.55819
Cracking output	kg	−265.38854	−265.38854
CH_4/process	kg	−57.98978	−57.98978
Waste air	kg	−588.08463	−588.08463
Net input–output balance			58.43324

CHEM FEEDSTOCK DISTRIB/NAPHTHA
Input

	Unit	Rate, unit	Rate, mass
Chem feedstock	kg	341.79861	341.79861

	Unit	Rate, unit	Rate, mass
Output			
Naphtha	kg	–341.11638	–341.11638
VOC	kg	–0.34112	–0.34112
Oils & greases	kg	–0.34112	–0.34112
Net input–output balance			0.00000
OIL REFINING/CHEM FEEDSTOCK			
Input			
Crude oil	kg	365.72451	365.72451
Salt water	kg	1298.83471	1298.83471
Process water	kg	136.71944	136.71944
NaOH/process	kg	0.23926	0.23926
Catalyst	kg	0.00342	0.00342
Process air	kg	329.32296	329.32296
Output			
Waste air	kg	–352.73616	–352.73616
Wastewater	kg	–1435.79683	–1435.79683
Oils & greases	kg	–0.00342	0.00000
Phenols	kg	–0.00034	0.00000
Sulfides	kg	–0.00051	0.00000
CO_2	kg	–68.35972	0.00000
SO_2	kg	–0.85450	0.00000
NO_x	kg	–0.11963	0.00000
VOC	kg	–0.25635	0.00000
CH_4	kg	–0.03418	0.00000
Solid waste/other industrial	kg	–0.51270	–0.51270
Chem feedstock	kg	–341.79861	–341.79861
Net input–output balance			0.00000
WASTE PAPER SURPLUS MANAG/CE			
Input			
Waste paper/collected/ce	kg	194296.15714	194296.15714
Output			
Waste paper/surplus/ce	kg	–194296.15714	–194296.15714
Net input–output balance			0.00000
WASTE INCIN/WASTE PAPER & BOARD			
Input			
Waste paper & board	kg	43174.54586	43174.54586
Electric power	kWh	1726.98183	0.00000
Process air	kg	345396.36692	345396.36692
Output			
CO_2	kg	–58285.63691	0.00000
CO	kg	–60.44436	0.00000
SO_2	kg ·	–56.12691	0.00000
NO_x	kg	–64.76182	0.00000
VOC	kg	–12.95236	0.00000
TSP	kg	–23.74600	0.00000

	Unit	Rate, unit	Rate, mass
HCl	kg	−224.50764	0.00000
Hf	g	−1899.68002	0.00000
Pb	g	−820.31637	0.00000
Zn	g	−2072.37820	0.00000
Cd	g	−47.49200	0.00000
Solid waste/incineration	kg	−2158.72729	−2158.72729
Heat	MJ	−447720.04061	0.00000
Waste air	kg	−386412.18549	−386412.18549
Electric power	kWh	−67520.19256	0.00000
Net input–output balance			0.00000
LANDFILL/WASTE PAPER & BOARD			
Input			
Waste paper & board	kg	122881.39976	122881.39976
Electric power	kWh	2457.62800	0.00000
Output			
Municipal waste/paper/landfill	kg	−12226.69928	−12226.69928
Leachate	kg	−614.40700	−614.40700
Organic compounds	kg	−110040.29348	−110040.29348
Net input–output balance			0.00000
WASTE PAPER SURPLUS MANAG/SC			
Input			
Waste paper/collected/sc	kg	3409.99824	3409.99824
Output			
Waste paper/surplus/sc	kg	−3409.99824	−3409.99824
Net input–output balance			0.00000
WASTE INCIN/OTHER MUNIC WASTE			
Input			
Solid waste/other municipal	kg	16866.92788	16866.92788
Electric power	kWh	674.67712	0.00000
Process air	kg	89394.71780	89394.71780
Output			
CO_2	kg	−60383.60183	0.00000
CO	kg	−23.61370	0.00000
SO_2	kg	−21.92701	0.00000
VOC	kg	−5.06008	0.00000
TSP	kg	−9.27681	0.00000
HCl	kg	−87.70802	0.00000
Hf	g	−742.14483	0.00000
Pb	g	−320.47163	0.00000
Zn	g	−809.61254	0.00000
Cd	g	−18.55362	0.00000
Solid waste/incineration	kg	−5060.07837	−5060.07837
Heat	MJ	−98334.18958	0.00000
Waste air	kg	−101201.56732	−101201.56732
Electric power	kWh	−14852.26706	0.00000
Net input–output balance			0.00000

	Unit	Rate, unit	Rate, mass
LANDFILL/OTHER MUNIC WASTE			
Input			
Solid waste/other municipal	kg	45603.17539	45603.17539
Electric power	kWh	912.06351	0.00000
Output			
CH_4	kg	−3283.42863	−3283.42863
Leachate	kg	−912.06351	−912.06351
Municipal waste/landfill	kg	−41407.68326	−41407.68326
Net input–output balance			0.00000
BIODEGRADATION/HARVESTING WASTES			
Input			
Harvesting wastes	kg	809125.76669	809125.76669
Output			
CH_4	kg	−99053.17636	−99053.17636
CO_2	kg	−338635.31587	−338635.31587
H_2O	kg	−371437.27446	−371437.27446
Net input–output balance			0.00000
WASTE PAPER SURPLUS DISP/CE			
Input			
Waste paper/surplus/ce	kg	11025.94736	11025.94736
Output			
Waste paper & board	kg	−11025.94736	−11025.94736
Net input–output balance			0.00000
LANDFILL/SLAG/INCINERATION			
Input			
Solid waste/incineration	kg	7218.80566	7218.80566
Electric power	kWh	144.37611	0.00000
Output			
Leachate	kg	−144.37611	−144.37611
Municipal waste/incin/landfill	kg	−7074.42955	−7074.42955
Net input–output balance			0.00000
DECOMPOSITION/ORGANIC COMPOUNDS			
Input			
Organic compounds	kg	110040.29348	110040.29348
Output			
CH_4	kg	−29344.44506	−29344.44506
CO_2	kg	−80695.84842	−80695.84842
Net input–output balance			0.00000
WASTE PAPER SURPLUS DISP/SC			
Input			
Waste paper/surplus/sc	kg	3409.99824	3409.99824
Output			
Waste paper & board	kg	−3409.99824	−3409.99824
Net input–output balance			0.00000

Appendix B: Energy Conversion Data for the Maximum Recycling Scenario (M)

	Unit	Rate, unit	Rate, mass
ELECTRIC GENERATION/W EUROPE			
Input			
Electric power/brown coal	kWh	101665.50927	0.00000
Electric power/hard coal	kWh	351208.12293	0.00000
Electric power/middle dist	kWh	2640.66258	0.00000
Electric power/heavy fuel oil	kWh	0.00000	0.00000
Electric power/natural gas	kWh	73938.55220	0.00000
Electric power/wood	kWh	0.00000	0.00000
Electric power/nuclear	kWh	392138.39289	0.00000
Electric power/hydro	kWh	271988.24558	0.00000
Output			
Electric power	kWh	−1320331.28921	0.00000
Net input–output balance			0.00000
ELECTRIC GENERATION/BROWN COAL			
Input			
Brown coal	kg	92177.38388	92177.38388
Process air	kg	792724.10034	792724.10034
Output			
Electric power/brown coal	kWh	−101665.50927	0.00000
SO_2	kg	−1987.91887	0.00000
NO_x	kg	−280.73548	0.00000
CO	kg	−29.11334	0.00000
CO_2	kg	−113957.79358	0.00000
VOC	kg	−2.83850	0.00000
Waste air	kg	−878441.98816	−878441.98816
Solid waste/other industrial	kg	−6456.91517	−6456.91517
Net input–output balance			2.58088
ELECTRIC GENERATION/HARD COAL			
Input			
Hard coal	kg	122632.34052	122632.34052
Process air	kg	3534574.51332	3534574.51332
Output			
Electric power/hard coal	kWh	−351208.12293	0.00000
SO_2	kg	−3231.99275	0.00000
NO_x	kg	−1144.73899	0.00000
CO	kg	−46.69488	0.00000
CO_2	kg	−358112.55538	0.00000
VOC	kg	−9.80573	0.00000
Waste air	kg	−3639673.54411	−3639673.54411
Solid waste/other industrial	kg	−17528.47817	−17528.47817
Net input–output balance			4.83157

	Unit	Rate, unit	Rate, mass
ELECTRIC GENERATION/HEAVY FUEL			
Input			
Heavy fuel oil	kg	0.00000	0.00000
Process air	kg	0.00000	0.00000
Output			
Electric power/heavy fuel oil	kWh	0.00000	0.00000
SO_2	kg	0.00000	0.00000
NO_x	kg	0.00000	0.00000
CO	kg	0.00000	0.00000
CO_2	kg	0.00000	0.00000
VOC	kg	0.00000	0.00000
Waste air	kg	0.00000	0.00000
Solid waste/other industrial	kg	0.00000	0.00000
Net input–output balance			0.00000
ELECTRIC GENERATION/HYDRO			
Input			
Hydro power	MJ	2781697.96632	0.00000
Output			
Electric power/hydro	kWh	−271988.24558	0.00000
Net input–output balance			0.00000
ELECTRIC GENERATION/MIDDLE DIST			
Input			
Middle distillate	kg	635.45356	635.45356
Process air	kg	10159.67920	10159.67920
Output			
Electric power/middle dist	kWh	−2640.66258	0.00000
SO_2	kg	−6.10893	0.00000
NO_x	kg	−4.52363	0.00000
CO	kg	−0.37810	0.00000
CO_2	kg	−1971.49467	0.00000
VOC	kg	−0.07373	0.00000
Waste air	kg	−10801.63028	−10801.63028
Solid waste/other industrial	kg	−0.96441	−0.96441
Net input–output balance			7.46193
ELECTRIC GENERATION/NATURAL GAS			
Input			
Natural gas	kg	15018.66409	15018.66409
Process air	kg	285355.75964	285355.75964
Output			
Electric power/natural gas	kWh	−73938.55220	0.00000
NO_x	kg	−113.57967	0.00000
CO	kg	−12.85525	0.00000
CO_2	kg	−38270.00644	0.00000
VOC	kg	−2.06436	0.00000

	Unit	Rate, unit	Rate, mass
Waste air	kg	–300373.68789	–300373.68789
Net input–output balance			0.73584
ELECTRIC GENERATION/NUCLEAR			
Input			
Nuclear fuel	MJ	4010506.29125	0.00000
Output			
Electric power/nuclear	kWh	–392138.39289	0.00000
Net input–output balance			0.00000
ELECTRIC GENERATION/WOOD			
Input			
Wood	kg	0.00000	0.00000
Process air	kg	0.00000	0.00000
Output			
Electric power/wood	kWh	0.00000	0.00000
CO_2	kg	0.00000	0.00000
CO	kg	0.00000	0.00000
SO_2	kg	0.00000	0.00000
NO_x	kg	0.00000	0.00000
VOC	kg	0.00000	0.00000
Waste air	kg	0.00000	0.00000
Solid waste/other industrial	kg	0.00000	0.00000
Net input–output balance			0.00000
HEAT GENERATION/W EUROPE			
Input			
Heat/brown coal	MJ	3801.37405	0.00000
Heat/hard coal	MJ	28948.92542	0.00000
Heat/middle distillate	MJ	21931.00411	0.00000
Heat/heavy fuel oil	MJ	56143.37051	0.00000
Heat/natural gas	MJ	84215.05577	0.00000
Heat/wood	MJ	9942.05519	0.00000
Heat/derived coal	MJ	60529.57133	0.00000
Heat/light fuel oil	MJ	26902.03170	0.00000
Output			
Heat	MJ	–292413.38808	0.00000
Net input–output balance			0.00000
HEAT GENERATION/BROWN COAL			
Input			
Brown coal	kg	421.25156	421.25156
Process air	kg	3622.75698	3622.75698
Output			
Heat/brown coal	MJ	–3801.37405	0.00000
CO_2	kg	–1066.14287	0.00000
CO	kg	–0.90758	0.00000
SO_2	kg	–3.53053	0.00000
NO_x	kg	–0.95034	0.00000

	Unit	Rate, unit	Rate, mass
VOC	kg	−0.21065	0.00000
TSP	kg	−1.18370	0.00000
HCl	kg	−0.58974	0.00000
Hf	g	−71.61276	0.00000
Pb	g	−3.53851	0.00000
Zn	g	−5.89750	0.00000
Cd	g	−0.09266	0.00000
Waste air	kg	−4014.48858	−4014.48858
Solid waste/other industrial	kg	−29.50817	−29.50817
Net input–output balance			0.01180

CO-GENERATION/BROWN COAL
Input

Brown coal	kg	0.00000	0.00000
Process air	kg	0.00000	0.00000

Output

Heat/co-gen/brown coal	MJ	0.00000	0.00000
CO_2	kg	0.00000	0.00000
CO	kg	0.00000	0.00000
SO_2	kg	0.00000	0.00000
NO_x	kg	0.00000	0.00000
VOC	kg	0.00000	0.00000
TSP	kg	0.00000	0.00000
HCl	kg	0.00000	0.00000
Hf	g	0.00000	0.00000
Pb	g	0.00000	0.00000
Zn	g	0.00000	0.00000
Cd	g	0.00000	0.00000
Electric power	kWh	0.00000	0.00000
Waste air	kg	0.00000	0.00000
Solid waste/other industrial	kg	0.00000	0.00000
Net input–output balance			0.00000

CO-GENERATION/DERIVED COAL
Input

Derived coal	kg	0.00000	0.00000
Process air	kg	0.00000	0.00000

Output

Heat/co-gen/derived coal	MJ	0.00000	0.00000
CO_2	kg	0.00000	0.00000
CO	kg	0.00000	0.00000
SO_2	kg	0.00000	0.00000
NO_x	kg	0.00000	0.00000
VOC	kg	0.00000	0.00000
TSP	kg	0.00000	0.00000
HCl	kg	0.00000	0.00000
Hf	g	0.00000	0.00000
Pb	g	0.00000	0.00000

	Unit	Rate, unit	Rate, mass
Zn	g	0.00000	0.00000
Cd	g	0.00000	0.00000
Electric power	kWh	0.00000	0.00000
Waste air	kg	0.00000	0.00000
Solid waste/other industrial	kg	0.00000	0.00000
Net input–output balance			0.00000
CO-GENERATION/HARD COAL			
Input			
Hard coal	kg	0.00000	0.00000
Process air	kg	0.00000	0.00000
Output			
Heat/co-gen/hard coal	MJ	0.00000	0.00000
CO_2	kg	0.00000	0.00000
CO	kg	0.00000	0.00000
SO_2	kg	0.00000	0.00000
NO_x	kg	0.00000	0.00000
VOC	kg	0.00000	0.00000
TSP	kg	0.00000	0.00000
HCl	kg	0.00000	0.00000
Hf	g	0.00000	0.00000
Pb	g	0.00000	0.00000
Zn	g	0.00000	0.00000
Cd	g	0.00000	0.00000
Electric power	kWh	0.00000	0.00000
Waste air	kg	0.00000	0.00000
Solid waste/other industrial	kg	0.00000	0.00000
Net input–output balance			0.00000
CO-GENERATION/HEAVY FUEL OIL			
Input			
Heavy fuel oil	kg	0.00000	0.00000
Process air	kg	0.00000	0.00000
Output			
Heat/co-gen/heavy fuel oil	MJ	0.00000	0.00000
CO_2	kg	0.00000	0.00000
CO	kg	0.00000	0.00000
SO_2	kg	0.00000	0.00000
NO_x	kg	0.00000	0.00000
VOC	kg	0.00000	0.00000
TSP	kg	0.00000	0.00000
HCl	kg	0.00000	0.00000
Hf	g	0.00000	0.00000
Pb	g	0.00000	0.00000
Zn	g	0.00000	0.00000
Cd	g	0.00000	0.00000
Electric power	kWh	0.00000	0.00000
Waste air	kg	0.00000	0.00000

	Unit	Rate, unit	Rate, mass
Solid waste/other industrial	kg	0.00000	0.00000
Net input–output balance			0.00000

CO-GENERATION/LIGHT FUEL OIL

Input

	Unit	Rate, unit	Rate, mass
Light fuel oil	kg	0.00000	0.00000
Process air	kg	0.00000	0.00000

Output

	Unit	Rate, unit	Rate, mass
Heat/co-gen/light fuel oil	MJ	0.00000	0.00000
CO_2	kg	0.00000	0.00000
CO	kg	0.00000	0.00000
NO_x	kg	0.00000	0.00000
VOC	kg	0.00000	0.00000
TSP	kg	0.00000	0.00000
HCl	kg	0.00000	0.00000
Hf	g	0.00000	0.00000
Pb	g	0.00000	0.00000
Zn	g	0.00000	0.00000
Cd	g	0.00000	0.00000
Electric power	kWh	0.00000	0.00000
Waste air	kg	0.00000	0.00000
Solid waste/other industrial	kg	0.00000	0.00000
Net input–output balance			0.00000

CO-GENERATION/MIDDLE DISTILLATE

Input

	Unit	Rate, unit	Rate, mass
Middle distillate	kg	0.00000	0.00000
Process air	kg	0.00000	0.00000

Output

	Unit	Rate, unit	Rate, mass
Heat/co-gen/middle dist	MJ	0.00000	0.00000
CO_2	kg	0.00000	0.00000
CO	kg	0.00000	0.00000
SO_2	kg	0.00000	0.00000
NO_x	kg	0.00000	0.00000
VOC	kg	0.00000	0.00000
TSP	kg	0.00000	0.00000
HCl	kg	0.00000	0.00000
Hf	g	0.00000	0.00000
Pb	g	0.00000	0.00000
Zn	g	0.00000	0.00000
Cd	g	0.00000	0.00000
Electric power	kWh	0.00000	0.00000
Waste air	kg	0.00000	0.00000
Solid waste/other industrial	kg	0.00000	0.00000
Net input–output balance			0.00000

CO-GENERATION/NATURAL GAS

Input

	Unit	Rate, unit	Rate, mass
Natural gas	kg	0.00000	0.00000

	Unit	Rate, unit	Rate, mass
Process air	kg	0.00000	0.00000
Output			
Heat/co-gen/natural gas	MJ	0.00000	0.00000
CO_2	kg	0.00000	0.00000
CO	kg	0.00000	0.00000
NO_x	kg	0.00000	0.00000
VOC	kg	0.00000	0.00000
Electric power	kWh	0.00000	0.00000
Waste air	kg	0.00000	0.00000
Net input–output balance			0.00000
CO-GENERATION/WOOD			
Input			
Wood	kg	0.00000	0.00000
Process air	kg	0.00000	0.00000
Output			
Heat/co-gen/wood	MJ	0.00000	0.00000
CO_2	kg	0.00000	0.00000
CO	kg	0.00000	0.00000
SO_2	kg	0.00000	0.00000
NO_x	kg	0.00000	0.00000
VOC	kg	0.00000	0.00000
TSP	kg	0.00000	0.00000
Electric power	kWh	0.00000	0.00000
Waste air	kg	0.00000	0.00000
Solid waste/other industrial	kg	0.00000	0.00000
Net input–output balance			0.00000
CO-GENERATION/HEAT/W EUROPE			
Input			
Heat/co-gen/brown coal	MJ	0.00000	0.00000
Heat/co-gen/hard coal	MJ	0.00000	0.00000
Heat/co-gen/middle dist	MJ	0.00000	0.00000
Heat/co-gen/heavy fuel oil	MJ	0.00000	0.00000
Heat/co-gen/natural gas	MJ	0.00000	0.00000
Heat/co-gen/wood	MJ	0.00000	0.00000
Heat/co-gen/derived coal	MJ	0.00000	0.00000
Heat/co-gen/light fuel oil	MJ	0.00000	0.00000
Output			
Heat/co-gen	MJ	0.00000	0.00000
Net input–output balance			0.00000
HEAT GENERATION/DERIVED COAL			
Input			
Derived coal	kg	2533.89031	2533.89031
Process air	kg	43582.80460	43582.80460
Output			
Heat/derived coal	MJ	–60529.57133	0.00000
CO_2	kg	–7761.40428	0.00000

	Unit	Rate, unit	Rate, mass
CO	kg	−7.25901	0.00000
SO$_2$	kg	−20.12608	0.00000
NO$_x$	kg	−10.59267	0.00000
VOC	kg	−1.29158	0.00000
TSP	kg	−7.25901	0.00000
HCl	kg	−3.61664	0.00000
Hf	g	−439.14434	0.00000
Pb	g	−21.69912	0.00000
Zn	g	−36.16491	0.00000
Cd	g	−0.56825	0.00000
Waste air	kg	−46025.17280	−46025.17280
Solid waste/other industrial	kg	−91.55098	−91.55098
Net input–output balance			0.02887

HEAT GENERATION/HARD COAL
Input

	Unit	Rate, unit	Rate, mass
Hard coal	kg	1235.44406	1235.44406
Process air	kg	35608.62571	35608.62571

Output

	Unit	Rate, unit	Rate, mass
Heat/hard coal	MJ	−28948.92542	0.00000
CO$_2$	kg	−3711.97596	0.00000
CO	kg	−3.47170	0.00000
SO$_2$	kg	−19.54052	0.00000
NO$_x$	kg	−8.32282	0.00000
VOC	kg	−0.61771	0.00000
TSP	kg	−3.47170	0.00000
HCl	kg	−1.72970	0.00000
Hf	g	−210.02555	0.00000
Pb	g	−10.37784	0.00000
Zn	g	−17.29626	0.00000
Cd	g	−0.27177	0.00000
Waste air	kg	−36667.43266	−36667.43266
Solid waste/other industrial	kg	−176.58845	−176.58845
Net input–output balance			0.04866

HEAT GENERATION/HEAVY FUEL OIL
Input

	Unit	Rate, unit	Rate, mass
Heavy fuel oil	kg	1724.30501	1724.30501
Process air	kg	27244.27233	27244.27233

Output

	Unit	Rate, unit	Rate, mass
Heat/heavy fuel oil	MJ	−56143.37051	0.00000
CO$_2$	kg	−5120.27539	0.00000
CO	kg	−0.42248	0.00000
SO$_2$	kg	−89.33814	0.00000
NO$_x$	kg	−11.93047	0.00000
VOC	kg	−0.55795	0.00000

	Unit	Rate, unit	Rate, mass
TSP	kg	−1.84364	0.00000
HCl	kg	−0.14878	0.00000
Hf	g	−14.88153	0.00000
Pb	g	−16.91038	0.00000
Zn	g	−6.42634	0.00000
Cd	g	−0.14176	0.00000
Waste air	kg	−28965.76843	−28965.76843
Solid waste/other industrial	kg	−2.58961	−2.58961
Net input–output balance			0.21930

HEAT GENERATION/LIGHT FUEL OIL

Input

Light fuel oil	kg	704.97987	704.97987
Process air	kg	11984.51885	11984.51885

Output

Heat/light fuel oil	MJ	−26902.03170	0.00000
CO_2	kg	−2367.37879	0.00000
CO	kg	−0.44423	0.00000
NO_x	kg	−2.35393	0.00000
VOC	kg	−0.29627	0.00000
TSP	kg	−0.01480	0.00000
HCl	kg	−0.00135	0.00000
Hf	g	−0.14796	0.00000
Pb	g	−0.59219	0.00000
Zn	g	−0.59219	0.00000
Cd	g	−0.05918	0.00000
Waste air	kg	−12688.67953	−12688.67953
Solid waste/other industrial	kg	−1.05760	−1.05760
Net input–output balance			0.23841

HEAT GENERATION/MIDDLE DIST

Input

Middle distillate	kg	646.77188	646.77188
Process air	kg	10340.63723	10340.63723

Output

Heat/middle distillate	MJ	−21990.24373	0.00000
CO_2	kg	−1935.14145	0.00000
CO	kg	−0.16163	0.00000
SO_2	kg	−4.89283	0.00000
NO_x	kg	−1.92415	0.00000
VOC	kg	−0.21331	0.00000
TSP	kg	−0.70507	0.00000
HCl	kg	−0.05691	0.00000
Hf	g	−5.69162	0.00000
Pb	g	−6.46762	0.00000
Zn	g	−2.45770	0.00000
Cd	g	−0.05443	0.00000

	Unit	Rate, unit	Rate, mass
Waste air	kg	−10994.02235	−10994.02235
Solid waste/other industrial	kg	−0.98160	−0.98160
Net input–output balance			7.59484

HEAT GENERATION/NATURAL GAS
Input

	Unit	Rate, unit	Rate, mass
Natural gas	kg	2106.39823	2106.39823
Process air	kg	40021.72722	40021.72722

Output

	Unit	Rate, unit	Rate, mass
Heat/natural gas	MJ	−84845.72232	0.00000
CO_2	kg	−5370.73422	0.00000
CO	kg	−0.71805	0.00000
NO_x	kg	−7.42400	0.00000
VOC	kg	−0.80179	0.00000
Waste air	kg	−42128.02227	−42128.02227
Net input–output balance			0.10317

HEAT GENERATION/WOOD
Input

	Unit	Rate, unit	Rate, mass
Wood	kg	776.72306	776.72306
Process air	kg	5903.09527	5903.09527

Output

	Unit	Rate, unit	Rate, mass
Heat/wood	MJ	−9942.05519	0.00000
CO_2	kg	−1422.95665	0.00000
CO	kg	−11.41783	0.00000
SO_2	kg	−0.05667	0.00000
NO_x	kg	−1.61558	0.00000
VOC	kg	−2.05055	0.00000
TSP	kg	−2.27587	0.00000
Waste air	kg	−6667.39077	−6667.39077
Solid waste/other industrial	kg	−12.42757	−12.42757
Net input–output balance			0.00000

FUEL DISTRIB/HEAVY FUEL OIL
Input

	Unit	Rate, unit	Rate, mass
Fuel	kg	49372.19142	49372.19142

Output

	Unit	Rate, unit	Rate, mass
Heavy fuel oil	kg	−48980.34863	−48980.34863
Oils & greases	kg	−293.88209	−293.88209
VOC	kg	−97.96070	−97.96070
Net input–output balance			0.00000

FUEL DISTRIB/LIGHT FUEL OIL
Input

	Unit	Rate, unit	Rate, mass
Fuel	kg	21975.95928	21975.95928

Output

	Unit	Rate, unit	Rate, mass
Light fuel oil	kg	−21888.40566	−21888.40566
VOC	kg	−87.55362	−87.55362
Net input–output balance			0.00000

	Unit	Rate, unit	Rate, mass
FUEL DISTRIB/MIDDLE DISTILLATE			
Input			
Fuel	kg	52825.47507	52825.47507
Output			
Middle distillate	kg	−52406.22527	−52406.22527
Oils & greases	kg	−104.81245	−104.81245
VOC	kg	−314.43735	−314.43735
Net input–output balance			0.00000
GAS PIPELINE DISTRIBUTION			
Input			
Natural gas/cleaned	m^3	111372.52314	82415.66712
Natural gas	kg	157.98691	157.98691
Process air	kg	45653.57012	45653.57012
Output			
Natural gas	kg	−78491.11154	−78491.11154
VOC	kg	−376.75734	0.00000
NO$_x$	kg	−4348.40758	0.00000
CH$_4$	kg	−1679.70979	0.00000
CO	kg	−549.43778	0.00000
Waste air	kg	−49736.67774	−49736.67774
Net input–output balance			0.56513
STEAM GENERATION/W EUROPE			
Input			
Steam/brown coal	MJ	104126.72517	37188.11612
Steam/hard coal	MJ	792965.06093	283201.80736
Steam/middle distillate	MJ	600731.10677	214546.82376
Steam/heavy fuel oil	MJ	1537871.63332	549239.86882
Steam/natural gas	MJ	2306807.44998	823859.80324
Steam/wood	MJ	272331.43507	97261.22677
Steam/derived coal	MJ	1658017.85468	592149.23358
Steam/light fuel oil	MJ	736896.82430	263177.43714
Output			
Steam	MJ	−8009748.09022	−2859480.06821
Net input–output balance			1144.24858
STEAM GENERATION/BROWN COAL			
Input			
Brown coal	kg	11538.86584	11538.86584
Process air	kg	99234.07067	99234.07067
Process water	kg	37188.85990	37188.85990
Output			
Steam/brown coal	MJ	−104126.72517	−37188.11612
CO$_2$	kg	−29203.64166	0.00000
CO	kg	−24.86026	0.00000
SO$_2$	kg	−96.70770	0.00000
NO$_x$	kg	−26.03168	0.00000

	Unit	Rate, unit	Rate, mass
VOC	kg	−5.76997	0.00000
TSP	kg	−32.42381	0.00000
HCl	kg	−16.15401	0.00000
Hf	g	−1961.60693	0.00000
Pb	g	−96.92636	0.00000
Zn	g	−161.54356	0.00000
Cd	g	−2.53809	0.00000
Waste air	kg	−109964.32970	−109964.32970
Solid waste/other industrial	kg	−808.28370	−808.28370
Net input–output balance			1.06688

STEAM GENERATION/DERIVED COAL
Input

	Unit	Rate, unit	Rate, mass
Derived coal	kg	69407.98157	69407.98157
Process air	kg	1193814.30581	1193814.30581
Process water	kg	592161.07680	592161.07680

Output

	Unit	Rate, unit	Rate, mass
Steam/derived coal	MJ	−1658017.85468	−592149.23358
CO_2	kg	−212599.33942	0.00000
CO	kg	−198.83779	0.00000
SO_2	kg	−551.29094	0.00000
NO_x	kg	−290.15312	0.00000
VOC	kg	−35.37878	0.00000
TSP	kg	−198.83779	0.00000
HCl	kg	−99.06657	0.00000
Hf	g	−12028.98254	0.00000
Pb	g	−594.37950	0.00000
Zn	g	−990.62422	0.00000
Cd	g	−15.56547	0.00000
Waste air	kg	−1260715.32625	−1260715.32625
Solid waste/other industrial	kg	−2507.75201	−2507.75201
Net input–output balance			11.05235

STEAM GENERATION/HARD COAL
Input

	Unit	Rate, unit	Rate, mass
Hard coal	kg	33841.11695	33841.11695
Process water	kg	283207.47151	283207.47151
Process air	kg	975386.67320	975386.67320

Output

	Unit	Rate, unit	Rate, mass
Steam/hard coal	MJ	−792965.06093	−283201.80736
CO_2	kg	−101677.94494	0.00000
CO	kg	−95.09633	0.00000
SO_2	kg	−535.25142	0.00000
NO_x	kg	−227.97746	0.00000
VOC	kg	−16.92029	0.00000
TSP	kg	−95.09633	0.00000
HCl	kg	−47.37966	0.00000
Hf	g	−5752.99165	0.00000

	Unit	Rate, unit	Rate, mass
Pb	g	−284.26846	0.00000
Zn	g	−473.77680	0.00000
Cd	g	−7.44436	0.00000
Waste air	kg	−1004389.37030	−1004389.37030
Solid waste/other industrial	kg	−4837.08687	−4837.08687
Net input–output balance			6.99712

STEAM GENERATION/HEAVY FUEL OIL

Input

Heavy fuel oil	kg	47231.93021	47231.93021
Process air	kg	746271.43346	746271.43346
Process water	kg	549250.85384	549250.85384

Output

Steam/heavy fuel oil	MJ	−1537871.63332	−549239.86882
CO_2	kg	−140253.89296	0.00000
CO	kg	−11.57248	0.00000
SO_2	kg	−2447.13824	0.00000
NO_x	kg	−326.79772	0.00000
VOC	kg	−15.28337	0.00000
TSP	kg	−50.50063	0.00000
HCl	kg	−4.07536	0.00000
Hf	g	−407.63287	0.00000
Pb	g	−463.20694	0.00000
Zn	g	−176.02940	0.00000
Cd	g	−3.88313	0.00000
Waste air	kg	−793426.42242	−793426.42242
Solid waste/other industrial	kg	−70.93433	−70.93433
Net input–output balance			16.99194

STEAM GENERATION/LIGHT FUEL OIL

Input

Light fuel oil	kg	19310.71362	19310.71362
Process water	kg	263182.70080	263182.70080
Process air	kg	328278.32402	328278.32402

Output

Steam/light fuel oil	MJ	−736896.82430	−263177.43714
CO_2	kg	−64846.92054	0.00000
CO	kg	−12.16838	0.00000
NO_x	kg	−64.47847	0.00000
VOC	kg	−8.11544	0.00000
TSP	kg	−0.40529	0.00000
HCl	kg	−0.03684	0.00000
Hf	g	−4.05293	0.00000
Pb	g	−16.22131	0.00000
Zn	g	−16.22131	0.00000
Cd	g	−1.62117	0.00000
Waste air	kg	−347566.59839	−347566.59839
Solid waste/other industrial	kg	−28.96962	−28.96962
Net input–output balance			1.26673

	Unit	Rate, unit	Rate, mass
STEAM GENERATION/MIDDLE DIST			
Input			
Middle distillate	kg	17668.56214	17668.56214
Process air	kg	282486.29382	282486.29382
Process water	kg	214551.11478	214551.11478
Output			
Steam/middle distillate	MJ	−600731.10677	−214546.82376
CO_2	kg	−52864.33740	0.00000
CO	kg	−4.41537	0.00000
SO_2	kg	−133.66267	0.00000
NO_x	kg	−52.56397	0.00000
VOC	kg	−5.82709	0.00000
TSP	kg	−19.26124	0.00000
HCl	kg	−1.55469	0.00000
Hf	g	−155.48423	0.00000
Pb	g	−176.68283	0.00000
Zn	g	−67.13951	0.00000
Cd	g	−1.48681	0.00000
Waste air	kg	−300335.51683	−300335.51683
Solid waste/other industrial	kg	−26.81544	−26.81544
Net input–output balance			203.18528
STEAM GENERATION/NATURAL GAS			
Input			
Natural gas	kg	57269.30002	57269.30002
Process water	kg	823876.28076	823876.28076
Process air	kg	1088121.07416	1088121.07416
Output			
Steam/natural gas	MJ	−2306807.44998	−823859.80324
CO_2	kg	−146020.91158	0.00000
CO	kg	−19.52251	0.00000
NO_x	kg	−201.84565	0.00000
VOC	kg	−21.79933	0.00000
Waste air	kg	−1145387.56910	−1145387.56910
Net input–output balance			19.28260
STEAM GENERATION/WOOD			
Input			
Wood	kg	21275.89336	21275.89336
Process air	kg	161696.78957	161696.78957
Process water	kg	97263.17203	97263.17203
Output			
Steam/wood	MJ	−272331.43507	−97261.22677
CO_2	kg	−38977.43664	0.00000
CO	kg	−312.75577	0.00000
SO_2	kg	−1.55229	0.00000
NO_x	kg	−44.25386	0.00000

	Unit	Rate, unit	Rate, mass
VOC	kg	−56.16836	0.00000
TSP	kg	−62.34021	0.00000
Waste air	kg	−182632.26864	−182632.26864
Solid waste/other industrial	kg	−340.41429	−340.41429
Net input–output balance			1.94526

COAL CLEANING/BROWN
Input

Coal/run-of-mine	kg	147865.81496	147865.81496
Electric power	kWh	856.99694	0.00000
Process water	kg	260327.13902	260327.13902
Process air	kg	112877.84748	112877.84748

Output

Brown coal	kg	−104130.85561	−104130.85561
TSP	kg	−2.57203	0.00000
NO_x	kg	−1.70775	0.00000
VOC	kg	−0.66644	0.00000
TDS	kg	−95.17560	0.00000
TSS	kg	−1.67651	0.00000
Al	kg	−0.11454	0.00000
NH_4	kg	−0.13537	0.00000
Sulfates	kg	−52.27369	0.00000
Solid waste/other industrial	kg	−32280.56524	−32280.56524
CO	kg	−0.46859	0.00000
Wastewater	kg	−260327.13902	−260327.13902
Waste air	kg	−124332.24160	−124332.24160
Net input–output balance			0.00000

COAL CLEANING/HARD
Input

Coal/run-of-mine	kg	367520.01472	367520.01472
Electric power	kWh	2130.06318	0.00000
Process water	kg	647042.27943	647042.27943
Process air	kg	280557.53236	280557.53236

Output

Hard coal	kg	−258816.91177	−258816.91177
TSP	kg	−6.39278	0.00000
NO_x	kg	−4.24460	0.00000
VOC	kg	−1.65643	0.00000
TDS	kg	−236.55866	0.00000
TSS	kg	−4.16695	0.00000
Al	kg	−0.28470	0.00000
NH_4	kg	−0.33646	0.00000
Sulfates	kg	−129.92609	0.00000
Solid waste/other industrial	kg	−80233.24265	−80233.24265
CO	kg	−1.16468	0.00000
Waste air	kg	−309027.39266	−309027.39266

	Unit	Rate, unit	Rate, mass
Wastewater	kg	–647042.27943	–647042.27943
Net input–output balance			0.00000

HARVESTING/WOOD/ENERGY
Input

	Unit	Rate, unit	Rate, mass
Wood biomass	kg	28184.38111	28184.38111
Light fuel oil	kg	3.46771	3.46771
Middle distillate	kg	46.40712	46.40712
Process air	kg	880.11624	880.11624

Output

	Unit	Rate, unit	Rate, mass
Wood	kg	–21984.95583	–21984.95583
Harvesting wastes	kg	–6199.46191	–6199.46191
Waste air	kg	–929.93760	–929.93760
CO	kg	–3.83075	0.00000
VOC	kg	–2.29068	0.00000
NO$_x$	kg	–2.56177	0.00000
TSP	kg	–0.19947	0.00000
Net input–output balance			0.01684

DERIVED COAL PRODUCTION
Input

	Unit	Rate, unit	Rate, mass
Electric power	kWh	3055.34151	0.00000
Heat	MJ	315073.12776	0.00000
Hard coal	kg	101127.50060	101127.50060
Process air	kg	114754.61061	114754.61061
Process water	kg	895085.96276	895085.96276

Output

	Unit	Rate, unit	Rate, mass
Derived coal	kg	–71721.63163	–71721.63163
SO$_2$	kg	–107.58245	0.00000
NO$_x$	kg	–50.20514	0.00000
CO	kg	–86.06596	0.00000
BOD5	kg	–45.90184	0.00000
Cd	g	–1.43443	0.00000
Cyanides	kg	–4.01641	0.00000
Coal tar	kg	–38012.46476	–38012.46476
Light fuel oil/impure	kg	–9538.97701	–9538.97701
TSP	kg	–71.72163	0.00000
VOC	kg	–21.51649	0.00000
Wastewater	kg	–3586.08158	–3586.08158
Derived coal/breeze	kg	–11690.62596	–11690.62596
Gas/derived coal production	kg	–17930.40791	–17930.40791
Water vapor	kg	–25102.57107	–25102.57107
Process water/used	kg	–884542.88291	–884542.88291
NH$_3$/process	kg	–48842.43114	–48842.43114
CO$_2$	kg	–1255.12855	0.00000
NH$_3$	kg	–0.35861	0.00000
Oils & greases	kg	–0.71722	0.00000
TDS	kg	–12.19268	0.00000
Net input–output balance			0.00000

	Unit	Rate, unit	Rate, mass
OIL REFINING/FUEL			
Input			
Crude oil	kg	132865.77957	132865.77957
Salt water	kg	471859.77791	471859.77791
Process water	kg	49669.45031	49669.45031
NaOH/process	kg	86.92154	86.92154
Catalyst	kg	1.24174	1.24174
Process air	kg	119641.28842	119641.28842
Output			
Waste air	kg	−128145.94005	−128145.94005
Wastewater	kg	−521617.39149	−521617.39149
Oils & greases	kg	−1.24174	0.00000
Phenols	kg	−0.12417	0.00000
Sulfides	kg	−0.18626	0.00000
CO_2	kg	−24834.72515	0.00000
SO_2	kg	−310.43406	0.00000
NO_x	kg	−43.46077	0.00000
VOC	kg	−93.13022	0.00000
CH_4	kg	−12.41736	0.00000
Solid waste/other industrial	kg	−186.26044	−186.26044
Fuel	kg	−124173.62576	−124173.62576
Net input–output balance			1.24174
NATURAL GAS WELL & PROCESSING			
Input			
Natural gas/crude	m³	115604.67902	85547.46247
Electric power	kWh	7069.92777	0.00000
Middle distillate	kg	1677.87295	1677.87295
Light fuel oil	kg	356.99917	356.99917
Natural gas	kg	3301.57162	3301.57162
Process air	kg	154700.88954	154700.88954
Output			
Natural gas/cleaned	m³	−111372.52314	−82415.66712
TDS	kg	−23.01179	−23.01179
VOC	kg	−37.86666	0.00000
CH_4	kg	−307.38816	0.00000
SO_2	kg	−249.47445	0.00000
Waste air	kg	−163160.74640	−163160.74640
Oils & greases	kg	−5.49289	−5.49289
Net input–output balance			20.12245
COAL MINING/SURFACE			
Input			
Coal seam	kg	402000.94715	402000.94715
Heat	MJ	314271.97122	0.00000
Process air	kg	745247.90972	745247.90972
Process water	kg	35561.62225	35561.62225

	Unit	Rate, unit	Rate, mass
Electric power	kWh	4947.70396	0.00000
Middle distillate	kg	371.07780	371.07780
Output			
Coal/run-of-mine	kg	–309231.49781	–309231.49781
Waste air	kg	–745247.90972	–745247.90972
Wastewater	kg	–35561.62225	–35561.62225
TSP	kg	–6.95771	0.00000
CH₄	kg	–507.22872	0.00000
VOC	kg	–2.78308	0.00000
SO₂	kg	–1.76262	0.00000
NOₓ	kg	–24.42929	0.00000
TDS	kg	–295.93454	0.00000
TSS	kg	–40.20009	0.00000
Fe/diss	kg	–3.71078	0.00000
Mn	g	–3246.93073	0.00000
Al	kg	–5.25694	0.00000
NH₄	kg	–3.71078	0.00000
Sulfates	kg	–134.20647	0.00000
CO	kg	–14.87404	0.00000
Solid waste/other industrial	kg	–92769.44934	–92769.44934
Net input–output balance			371.07780

COAL MINING/UNDERGROUND

	Unit	Rate, unit	Rate, mass
Input			
Electric power	kWh	3916.93231	0.00000
Process water	kg	49477.03965	49477.03965
Coal seam	kg	268000.63143	268000.63143
Process air	kg	496831.93981	496831.93981
Output			
NH₄	kg	–5.56617	0.00000
CH₄	kg	–1483.12477	0.00000
Chlorides	g	–47415.49633	0.00000
Coal/run-of-mine	kg	–206154.33187	–206154.33187
TDS	kg	–2195.54363	0.00000
TSS	kg	–105.13871	0.00000
Fe/susp	kg	–39.16932	0.00000
Fe/diss	kg	–123.69260	0.00000
Mn	g	–3360.31561	0.00000
Al	kg	–19.99697	0.00000
Sulfates	kg	–1096.74105	0.00000
Wastewater	kg	–49477.03965	–49477.03965
Waste air	kg	–496831.93981	–496831.93981
Solid waste/other industrial	kg	–61846.29956	–61846.29956
TSP	kg	–4.12309	0.00000
VOC	kg	–20.61543	0.00000
SO₂	kg	–20.61543	0.00000
NOₓ	kg	–16.49235	0.00000

	Unit	Rate, unit	Rate, mass
CO	kg	−10.30772	0.00000
Net input–output balance			0.00000
WOOD BIOMASS FORMATION			
Input			
CO_2	kg	22360.64245	22360.64245
H_2O	kg	22086.12657	22086.12657
Output			
Wood biomass	kg	−28184.38111	−28184.38111
O_2	kg	−16262.38790	−16262.38790
Net input–output balance			0.00000
OIL DISTRIBUTION			
Input			
Oil/extracted	kg	132865.77957	132865.77957
Natural gas	kg	420.81250	420.81250
Process air	kg	8211.10518	8211.10518
Output			
Crude oil	kg	−132865.77957	−132865.77957
Waste air	kg	−8415.71848	−8415.71848
Net input–output balance			216.19920
NaOH/DIAPHRAGM PROCESS			
Input			
NaCl/process	kg	140.81289	140.81289
H_2SO_4/process	kg	4.57468	4.57468
HCl/process	kg	4.34608	4.34608
Cooling water	l	12892.20252	12892.20252
Steam	MJ	23.49316	8.38706
Process water	kg	687.54937	687.54937
$CaCl_2$/process	kg	0.99438	0.99438
Electric power	kWh	0.23208	0.00000
Na_2CO_3/process	kg	6.08451	6.08451
Heat	MJ	761.43267	0.00000
Output			
NaOH/process	kg	−86.92154	−86.92154
H_2SO_4/process	kg	−6.08451	−6.08451
H_2/process	kg	−2.28777	−2.28777
Cooling water	l	−12892.20252	−12892.20252
Wastewater	kg	−657.99604	−657.99604
Solid waste/other industrial	kg	−12.19944	−12.19944
Cl_2	kg	−0.00610	−0.00610
H_2	kg	−0.28597	−0.28597
TDS	kg	−4.19396	0.00000
NaOH	kg	−0.08344	0.00000
Chlorides	g	−76.49095	0.00000
Sulfates	kg	−0.04194	0.00000
Asbestos	g	−0.76230	0.00000

	Unit	Rate, unit	Rate, mass
Cl$_2$/process	kg	−76.49095	−76.49095
Net input–output balance			10.47664
OIL WELL			
Input			
Heavy fuel oil	kg	217.44933	217.44933
Natural gas	kg	449.63255	449.63255
Crude oil/reservoir	t	137.48951	137489.50870
Electric power	kWh	29575.92253	0.00000
Salt water	kg	717475.20980	717475.20980
Light fuel oil	kg	19.52595	19.52595
Process air	kg	55751.80976	55751.80976
Output			
Oil/extracted	kg	−132865.77957	−132865.77957
Solid waste/other industrial	kg	−79.71947	−79.71947
VOC	kg	−13.28658	0.00000
TDS	kg	−1680.75211	0.00000
CH$_4$	kg	−13.28658	0.00000
Wastewater	kg	−719177.22031	−719177.22031
Waste air	kg	−59280.72487	−59280.72487
Oils & greases	kg	−21.25852	0.00000
Net input–output balance			0.30812
SALT MINING & PREPARATION			
Input			
Heat	MJ	35.36008	0.00000
Electric power	kWh	14.08410	0.00000
Salt rock	kg	169.50365	169.50365
Output			
NaCl/process	kg	−149.83086	−149.83086
Solid waste/other industrial	kg	−19.61286	−19.61286
TSP	kg	−0.04495	−0.04495
Acids	g	−11.98647	−0.01199
Heavy metals	kg	−0.00300	−0.00300
Net input–output balance			0.00000
H$_2$SO$_4$ PRODUCTION			
Input			
S$_2$/process	kg	0.00000	0.00000
Catalyst	kg	0.00000	0.00000
Cooling water	l	0.00000	0.00000
Process water	kg	0.00000	0.00000
Output			
H$_2$SO$_4$/process	kg	0.00000	0.00000
Steam	MJ	0.00000	0.00000
Cooling water	l	0.00000	0.00000
SO$_2$	kg	0.00000	0.00000

	Unit	Rate, unit	Rate, mass
H_2SO_4	kg	0.00000	0.00000
Sulfates	kg	0.00000	0.00000
Net input–output balance			0.00000
Na_2CO_3/SOLVAY PROCESS			
Input			
NaCl/process	kg	9.12676	9.12676
$CaCO_3$/process	kg	7.30141	7.30141
NH_3/process	kg	2.47031	2.47031
Derived coal	kg	0.60845	0.60845
Process air	kg	40.82705	40.82705
Process water	kg	29.38817	29.38817
Electric power	kWh	0.18722	0.00000
Heat	MJ	22.85341	0.00000
CO_2/process	kg	7.90986	7.90986
Output			
Na_2CO_3/process	kg	−6.08451	−6.08451
CO_2/process	kg	−13.38592	−13.38592
NH_3/process	kg	−2.45814	−2.45814
Wastewater	kg	−34.87640	−34.87640
Waste air	kg	−40.82705	−40.82705
NH_3	kg	−0.01217	0.00000
TSP	kg	−0.15211	0.00000
Net input–output balance			0.00000
LIMESTONE MINING & PREPARATION			
Input			
Limestone rock	kg	8.27834	8.27834
Process air	kg	7.63070	7.63070
Process water	kg	7.66648	7.66648
Electric power	kWh	0.41895	0.00000
Heat	MJ	0.83382	0.00000
Explosive	kg	0.00730	0.00730
Output			
$CaCO_3$/process	kg	−7.30141	−7.30141
Wastewater	kg	−8.35646	−8.35646
Waste air	kg	−7.91838	−7.91838
TSP	kg	−0.20502	0.00000
Net input–output balance			0.00657
NH_3 PRODUCTION			
Input			
Natural gas	kg	0.00000	0.00000
Process air	kg	0.00000	0.00000
Process water	kg	0.00000	0.00000
Cooling water	l	0.00000	0.00000

	Unit	Rate, unit	Rate, mass
Electric power	kWh	0.00000	0.00000
Heat	MJ	0.00000	0.00000
Output			
NH_3/process	kg	0.00000	0.00000
CO_2/process	kg	0.00000	0.00000
Waste air	kg	0.00000	0.00000
Wastewater	kg	0.00000	0.00000
Steam	MJ	0.00000	0.00000
NO_x	kg	0.00000	0.00000
NH_3/air	kg	0.00000	0.00000
CO	kg	0.00000	0.00000
H_2	kg	0.00000	0.00000
NH_3	kg	0.00000	0.00000
Mono-ethyl amine	kg	0.00000	0.00000
Cooling water	l	0.00000	0.00000
Net input–output balance			0.00000

BIODEGRADATION/HARVESTING WASTES

Input

Harvesting wastes	kg	6199.46191	6199.46191
Output			
CH_4	kg	–758.93813	–758.93813
CO_2	kg	–2594.59880	–2594.59880
H_2O	kg	–2845.92499	–2845.92499
Net input–output balance			0.00000

Appendix C: Transportation Process Data

	Distance (km)	Relative mass (%)
Product: Heavy fuel oil		
Rail/electric	215	60
Rail/diesel	215	15
16-t truck/diesel/highway	90	5
Ship/inland	147	20
Product: Wood		
Rail/electric	215	60
Rail/diesel	215	15
16-t truck/diesel/highway	90	5
Ship/inland	147	20
Product: Crude oil		
Pipeline	670	100
Ship/seagoing	5942	100
Pipeline	200	60
Product: Brown coal		
Rail/electric	203	75
Rail/diesel	203	25
Ship/inland	46	100
Ship/seagoing	8	100
Product: Hard coal		
Rail/electric	380	75
Rail/diesel	380	25
Ship/inland	71	100
Ship/seagoing	1816	100
Product: NaCl		
Rail/electric	269	75
Rail/diesel	269	25
Ship/inland	37	100
Ship/seagoing	102	100
Product: Derived coal		
Rail/electric	113	75
Rail/diesel	113	25
Ship/inland	13	100
Ship/seagoing	128	100
Product: Waste paper/surplus/sc		
Rail/electric	312	16
Rail/diesel	312	6
Ship/inland	165	5
ld[a] truck/gasoline/highway	144	3
hd[b] truck/diesel/highway	144	7

[a]low duty less than 8-tonne load; [b]heavy duty less than 16-tonne load.

	Distance (km)	Relative mass (%)
16-t truck/diesel/highway	144	7
ld truck/gasoline/rural	144	13
hd truck/diesel/rural	144	12
16-t truck/diesel/rural	144	14
ld truck/gasoline/urban	11	8
hd truck/diesel/urban	11	5
16-t truck/diesel/urban	11	4
Product: Waste paper/collected/ce		
Rail/electric	312	16
Rail/diesel	312	6
Ship/inland	165	5
ld truck/gasoline/highway	144	3
hd truck/diesel/highway	144	7
16-t truck/diesel/highway	144	7
ld truck/gasoline/rural	144	13
hd truck/diesel/rural	144	12
16-t truck/diesel/rural	144	14
ld truck/gasoline/urban	11	8
hd truck/diesel/urban	11	5
16-t truck/diesel/urban	11	4
Product: Waste paper/imported/sc/ce		
Rail/electric	1351	75
Rail/diesel	1351	25
Ship/seagoing	398	100
Product: Waste paper/imported/ow/ce		
Rail/electric	1351	75
Rail/diesel	1351	25
Ship/seagoing	5800	100
Product: Waste paper/collected/sc		
Rail/electric	312	16
Rail/diesel	312	6
Ship/inland	165	5
ld truck/gasoline/highway	144	3
hd truck/diesel/highway	144	7
16-t truck/diesel/highway	144	7
ld truck/gasoline/rural	144	13
hd truck/diesel/rural	144	12
16-t truck/diesel/rural	144	14
ld truck/gasoline/urban	11	8
hd truck/diesel/urban	11	5
16-t truck/diesel/urban	11	4
Product: Waste paper/imported/ce/sc		
Rail/electric	1351	75
Rail/diesel	1351	25
Ship/seagoing	398	100

	Distance (km)	Relative mass (%)
Product: Waste paper/imported/ow/sc		
Rail/electric	1351	75
Rail/diesel	1351	25
Ship/seagoing	6800	100
Product: Waste paper/surplus/sc		
Rail/electric	312	16
Rail/diesel	312	6
Ship/inland	165	5
ld truck/gasoline/highway	144	3
hd truck/diesel/highway	144	7
16-t truck/diesel/highway	144	7
ld truck/gasoline/rural	144	13
hd truck/diesel/rural	144	12
16-t truck/diesel/rural	144	14
ld truck/gasoline/urban	11	8
hd truck/diesel/urban	11	5
16-t truck/diesel/urban	11	4
Product: Logs		
Rail/electric	300	19
16-t truck/diesel/rural	115	20
16-t truck/diesel/highway	115	20
Ship/seagoing	250	2
Ship/inland	300	39
Product: Foods, including grains		
Rail/electric	311	5
Rail/diesel	311	2
Ship/inland	208	3
ld truck/gasoline/highway	134	3
hd truck/diesel/highway	134	9
16-t truck/diesel/highway	134	8
ld truck/gasoline/rural	134	15
hd truck/diesel/rural	134	15
16-t truck/diesel/rural	134	16
ld truck/gasoline/urban	17	12
hd truck/diesel/urban	17	6
16-t truck/diesel/urban	17	6
Product: Sand, clay		
Rail/electric	172	4
Rail/diesel	172	1
Ship/inland	112	17
ld truck/gasoline/highway	81	1
hd truck/diesel/highway	81	2
16-t truck/diesel/highway	81	1
ld truck/gasoline/rural	81	3
hd truck/diesel/rural	81	2

	Distance (km)	Relative mass (%)
16-t truck/diesel/rural	81	3
ld truck/gasoline/urban	10	32
hd truck/diesel/urban	10	18
16-t truck/diesel/urban	10	17
Product: Stones, minerals, salt		
Rail/electric	118	6
Rail/diesel	118	2
Ship/inland	136	9
ld truck/gasoline/highway	106	1
hd truck/diesel/highway	106	3
16-t truck/diesel/highway	106	2
ld truck/gasoline/rural	106	5
hd truck/diesel/rural	106	4
16-t truck/diesel/rural	106	5
ld truck/gasoline/urban	4	31
hd truck/diesel/urban	4	17
16-t truck/diesel/urban	4	16
Product: Ore, metal wastes		
Rail/electric	123	42
Rail/diesel	123	14
Ship/inland	81	27
ld truck/gasoline/highway	115	1
hd truck/diesel/highway	115	2
16-t truck/diesel/highway	115	1
ld truck/gasoline/rural	115	3
hd truck/diesel/rural	115	3
16-t truck/diesel/rural	115	3
ld truck/gasoline/urban	11	3
hd truck/diesel/urban	11	1
16-t truck/diesel/urban	11	1
Product: Solid fossil fuels		
Rail/electric	102	46
Rail/diesel	102	15
Ship/inland	176	24
ld truck/gasoline/highway	119	0
hd truck/diesel/highway	119	1
16-t truck/diesel/highway	119	1
ld truck/gasoline/rural	119	2
hd truck/diesel/rural	119	2
ld truck/gasoline/urban	8	4
hd truck/diesel/urban	8	2
16-t truck/diesel/urban	8	2
16-t truck/diesel/rural	119	2
Product: Oils, petroleum products		
Rail/electric	156	6

	Distance (km)	Relative mass (%)
Rail/diesel	156	2
Ship/inland	148	10
ld truck/gasoline/highway	82	0
hd truck/diesel/highway	82	1
16-t truck/diesel/highway	82	1
ld truck/gasoline/rural	82	2
hd truck/diesel/rural	82	2
16-t truck/diesel/rural	82	2
ld truck/gasoline/urban	17	9
hd truck/diesel/urban	17	5
16-t truck/diesel/urban	17	4
Pipeline	144	56
Product: Fertilizers		
Rail/electric	243	43
Rail/diesel	243	14
Ship/inland	197	20
ld truck/gasoline/highway	119	1
hd truck/diesel/highway	119	2
16-t truck/diesel/highway	119	2
ld truck/gasoline/rural	119	4
hd truck/diesel/rural	119	4
16-t truck/diesel/rural	119	4
ld truck/gasoline/urban	14	3
hd truck/diesel/urban	14	2
16-t truck/diesel/urban	14	1
Product: Mineral/construction products, glass		
Rail/electric	175	4
Rail/diesel	175	1
Ship/inland	155	1
ld truck/gasoline/highway	106	3
hd truck/diesel/highway	106	7
16-t truck/diesel/highway	106	6
ld truck/gasoline/rural	106	12
hd truck/diesel/rural	106	11
16-t truck/diesel/rural	106	13
ld truck/gasoline/urban	16	20
hd truck/diesel/urban	16	11
16-t truck/diesel/urban	16	11
Product: Iron, steel, including intermediates		
Rail/electric	145	31
Rail/diesel	145	10
Ship/inland	154	9
ld truck/gasoline/highway	147	2
hd truck/diesel/highway	147	5
16-t truck/diesel/highway	147	5

	Distance (km)	Relative mass (%)
ld truck/gasoline/rural	147	9
hd truck/diesel/rural	147	9
16-t truck/diesel/rural	147	10
ld truck/gasoline/urban	19	5
hd truck/diesel/urban	19	3
16-t truck/diesel/urban	19	3
Product: Machinery		
Rail/electric	386	7
Rail/diesel	386	2
Ship/inland	187	1
ld truck/gasoline/highway	164	4
hd truck/diesel/highway	164	10
16-t truck/diesel/highway	164	9
hd truck/diesel/rural	164	17
16-t truck/diesel/rural	164	19
ld truck/gasoline/urban	14	6
hd truck/diesel/urban	14	3
16-t truck/diesel/urban	14	3
ld truck/gasoline/rural	164	18
Product: Other goods, including chemical products		
Rail/electric	312	16
Rail/diesel	312	6
Ship/inland	165	5
ld truck/gasoline/highway	144	3
hd truck/diesel/highway	144	7
16-t truck/diesel/highway	144	7
ld truck/gasoline/rural	144	13
hd truck/diesel/rural	144	12
16-t truck/diesel/rural	144	14
ld truck/gasoline/urban	11	8
hd truck/diesel/urban	11	5
16-t truck/diesel/urban	11	4
Product: Light fuel oil		
16-t truck/diesel/highway	225	2
16-t truck/diesel/rural	225	4
16-t truck/diesel/urban	47	15
Rail/diesel	100	4
Rail/electric	100	13
16-t truck/diesel/highway	112	7
16-t truck/diesel/rural	112	16
16-t truck/diesel/urban	23	57
Ship/inland	200	35
Pipeline	100	27
Product: Middle distillate		
16-t truck/diesel/highway	225	2

	Distance (km)	Relative mass (%)
16-t truck/diesel/rural	225	4
16-t truck/diesel/urban	47	15
Rail/diesel	100	4
Rail/electric	100	13
16-t truck/diesel/highway	112	7
16-t truck/diesel/rural	112	16
16-t truck/diesel/urban	23	57
Ship/inland	200	35
Pipeline	100	27
Product: Waste paper, board		
Rail/electric	312	16
Rail/diesel	312	6
Ship/inland	165	5
ld truck/gasoline/highway	144	3
hd truck/diesel/highway	144	7
16-t truck/diesel/highway	144	7
ld truck/gasoline/rural	144	13
hd truck/diesel/rural	144	12
16-t truck/diesel/rural	144	14
ld truck/gasoline/urban	11	8
hd truck/diesel/urban	11	5
16-t truck/diesel/urban	11	4

References

Adams, D.M., and Haynes, R.W., 1991, Softwood Timber Supply and the Future of the Southern Forest Economy, *Southern Journal of Applied Forestry* **15**(1).

Benn Publications, 1987, *Paper European Databook 1987*, Benn Publications Limited, Kent, UK.

CEC, 1985, *Liquid Food Container Directive: EEC Directive 85/339*, Commission of the European Communities, Brussels, Belgium.

CTS-Engineering, 1991, Research contract report, unpublished.

Ebeling, K., 1991, *Effect of Waste Paper Recovery Rate on Future Fiber Needs and Paper Quality in Western Europe: A Mathematical Simulation Analysis*, unpublished.

Environmental News, 1991, No More Waste Packaging on Landfill Sites, *Environmental News from the Netherlands*, No. 3, Ministry of Housing, Physical Planning, and Environment, Leidschendam, The Netherlands.

FAO, 1990, *Yearbook 1988, Forest Products*, Food and Agriculture Organization of the United Nations, Rome, Italy.

FAO, 1991a, *FAO Yearbook, Forest Products*, Food and Agriculture Organization of the United Nations, Rome, Italy.

FAO, 1991b, *Waste Paper Data 1988–1990*, Food and Agriculture Organization of the United Nations, Rome, Italy.

Jaakko Pöyry, 1990, *Recycled Fibre: An Under-utilized Opportunity*, Jaako Pöyry Consulting, Vantaa, Finland.

Jaakko Pöyry, 1992, *World Paper Market up to 2005*, Jaakko Pöyry Consulting, Vantaa, Finland.

Judt, M., 1991, Personal communications.

Kyrklund, B.,1991, *Fibre Balance in Some European Countries*, unpublished.

Lübkert, B., Virtanen, Y., Mühlberger, M., Ingman, J., Vallance, B., and Alber, S., 1991, *Life-Cycle Analysis: IDEA – An International Database for Ecoprofile Analysis*, Working Paper WP-91-30, International Institute for Applied Systems Analysis, Laxenburg, Austria.

Meadows, D.H., Meadows, D.L., Randers, J., and Behrens, W.W., 1972, *The Limits to Growth*, Universe Books, A Potomac Associate Book, New York, NY, USA.

Mesarovic, M.D., and Pestel, E., 1974, *Mankind at the Turning Point*, Dutton, New York, NY, USA.

Nilsson, S., Sallnäs, O., and Duinker, P., 1992, *Future Forest Resources of Western and Eastern Europe*, Parthenon Publishing Group, Lancaster, UK.

OECD, 1989, *The Pulp and Paper Industry in the OECD Member Countries 1986*, Organisation for Economic Co-operation and Development, Paris, France.

Sandberg, E.,1990, Personal communications.

Schwab, F., 1991, Personal communications.

Smith, H., 1963, *Power from Geo-thermal Steam at Wairakey, New Zealand*, Paper presented at the 6th World Energy Conference, Melbourne, Australia, October 20–22, 1962.

Srivastava *et al.*, 1990, *Methanogenic Gasification of Wood*, Institute of Gas Technology, Chicago, IL, USA.

UN, 1986, *European Timber Trends and Prospects to the Year 2000 and Beyond*, United Nations, New York, NY, USA.

UN, 1991, *Survey of Medium-term Trends for Wood Raw Material, Notably Pulpwood, and Wood for Energy*, United Nations, New York, NY, USA.

Wiseman, C.A., 1990, *US Wastepaper Recycling Policies: Issues and Effects*, Discussion paper ENR 90-14, Resources for the Future, Washington, DC, USA.

Index

161

sorting, 100
starch, 26, 28
steam, 51, 59, 60, 67, 95, 97
storing, 26, 27
sulfate principle, 23
sulfite principle, 23
supply, 3, 9, 13, 40, 41, 46, 51, 52, 55
supply balance, 41
suspended solids, 19 64, 80
sustainability, 3, 89, 91
sustainable harvest level, 89

techno-economical aspect, 6
thermo-mechanical bleached, 49
thermomechanical pulp, 52
thickening, 26, 27
thinnings, 14, 21, 90, 92, 101, 102
total valuation, 99
trade balance, 11
trade patterns, 103
transportation, 11, 13, 18, 19, 27, 30, 39, 40, 45, 48, 50–52, 60, 61
trends, 13, 59–62, 64, 65, 93, 95
trimming, 29
TSSs, 19
        see also suspended solids

unbleached, 49, 95
uncleaned pulp, 27

vital forests, 90, 102
vitality of the forest resources, 14

waste-handling processes, 3
waste heat, 19
waste management, 19, 29, 39, 40
waste paper, 3, 9–11, 13, 15–19, 26, 28–32, 40, 42, 44–46, 48, 50, 52, 53, 55, 59–62, 64, 89, 90, 93, 99, 100, 102
waste paper balance, 46
waste paper categories, 44
waste paper collection, 10, 41

waste paper feed, 26, 27
waste paper fibers, 29, 31, 35
waste paper grades, 18, 28, 102
waste paper pulp, 49, 95, 97
waste paper qualities, 90, 102
waste paper raw material balance, 41
waste paper recovery, 15
waste paper utilization, 27
waste production, 44
waste sludge, 102
waste streams, 1, 2, 40
waste-management resources, 1
waste-management systems, 30
waste-treatment capacity, 2
wastewater discharge, 1
wastewater purifiers, 64
water emissions, 6, 19, 64
water glass, 27
water pollution, 5
weight of papers, 16, 37, 48
wood balance, 2, 3, 14, 16, 87–90, 101, 102
wood-based energy, 59
wood biomass, 23, 95
wood consumption, 87–90, 89, 101, 102
wood material, 15, 23
wood pulp production, 49
wood pulp supply, 14
wood supply, 87, 89, 101
wood supply/demand balance, 102
worksheet model, 40, 42
worksheet program, 40, 52

yield, 16, 23, 37, 46, 50
yield factors, 28

zero recycling, 59, 60, 62, 64, 65, 95
zero recycling scenario, 18, 49, 59–61, 64, 89, 102

The International Institute for Applied Systems Analysis

is a nongovernmental research institution, bringing together scientists from around the world to work on problems of common concern. Situated in Laxenburg, Austria, IIASA was founded in October 1972 by the academies of science and equivalent organizations of twelve countries. Its founders gave IIASA a unique position outside national, disciplinary, and institutional boundaries so that it might take the broadest possible view in pursuing its objectives:

To promote international cooperation in solving problems arising from social, economic, technological, and environmental change;

To create a network of institutions in the national member organization countries and elsewhere for joint scientific research;

To develop and formalize systems analysis and the sciences contributing to it and promote the use of analytical techniques needed to evaluate and address complex problems;

To inform policy advisers and decision makers about the potential application of the Institute's work to such problems.

The Institute now has national member organizaitons in the following countries:

Austria
The Austrian Academy of Sciences

Bulgaria
The National Committee for Applied Systems Analysis and Management

Canada
The Canadian Committee for IIASA

Czech Republic and Slovak Republic
The Joint Interim Committee for IIASA of the Czech Republic and the Slovak Republic

Finland
The Finnish Committee for IIASA

France
The French Association for the Development of Systems Analysis

Germany
The Association for the Advancement of IIASA

Hungary
The Hungarian Committee for Applied Systems Analysis

Italy
The National Research Council (CNR) and the National Commission for Nuclear and Alternative Energy Sources (ENEA)

Japan
The Japan Committee for IIASA

Netherlands
The Netherlands Organization for Scientific Research (NWO)

Poland
The Polish Academy of Sciences

Russia
The Russian Academy of Sciences

Sweden
The Swedish Council for Planning and Coordination of Research (FRN)

United States of America
The American Academy of Arts and Sciences